German Artillery

1914–1918

Wolfgang Fleischer

Pen & Sword
MILITARY

First published in Great Britain in 2015 by
Pen & Sword Military
an imprint of
Pen & Sword Books Ltd
47 Church Street
Barnsley
South Yorkshire
S70 2AS

Copyright © Wolfgang Fleischer 2015

ISBN 978 1 47382 398 3

Typeset in Ehrhardt by
Mac Style Ltd, Bridlington, East Yorkshire
Printed and bound in Malta by Gutenberg Press

Pen & Sword Books Ltd incorporates the imprints of Pen & Sword Archaeology, Atlas,
Aviation, Battleground, Discovery, Family History, History, Maritime, Military, Naval,
Politics, Railways, Select, Transport, True Crime, and Fiction, Frontline Books, Leo Cooper,
Praetorian Press, Seaforth Publishing and Wharncliffe.

For a complete list of Pen & Sword titles please contact
PEN & SWORD BOOKS LIMITED
47 Church Street, Barnsley, South Yorkshire, S70 2AS, England
E-mail: enquiries@pen-and-sword.co.uk
Website: www.pen-and-sword.co.uk

Photo credits: Unless otherwise indicated, all photos come from the author's archive.

Contents

Foreword

During the First World War 65 per cent of the combatants were killed by artillery fire. In places such as Verdun, 1916, where the artillery was particularly prominent, the rate was as much as 75 per cent. By comparison, in the Franco-Geman War of 1870/71, artillery caused only 7.5 per cent of the deaths. There is no more impressive manner of demonstrating how artillery had come to the forefront by the beginning of the twentieth century – unless one shows the number of guns and the enormous increase in the numbers of shells fired.

The battles of the First World War were marked by high concentrations of artillery in a small area and by artillery bombardments which lasted for days. This volume of the 'Typenkompass' series is devoted to the artillery of the German Army and the First World War coastal artillery operated by the Imperial German Navy from 1888 onwards. In the selection of over 100 guns, the intention is to show the development of the artillery arm of the German Army up to the outbreak of war and the changes occurring during its course. The lack of guns and ammunition led to the use of both captured and imported artillery. Captured weapons were tied to an interesting trend in development. From the mid-19th century the export of armaments had been constantly rising. In some areas this resulted in an assimilation of types (e.g. the 3.7-cm Hotchkiss revolver cannon and the Krupp 28-cm coastal gun). In its turn that made it easier to use captured material: these stocks of guns and ammunition represented a valuable resource for the German artillery.

It was the author's initial plan to present the whole diversity of gun types in all their varying carriages, mounts and uses. In view of the numerous calibres and gun models, however, the author was forced to abandon comprehensive coverage, technical descriptions and the ammunition. The same also goes for interesting subjects such as the composition and tactical use of the German artillery between 1914 and 1918.

January 2013   *Wolfgang Fleischer*

The battlefield of Verdun, photograph taken in 1916.

When the Great War broke out in August 1914, the field and *Fuss-artillerie* (*see Glossary) of the German Army had over 1,270 batteries with around 8,700 guns. Since the turn of the nineteenth century – and not without opposition from within its own ranks – modern guns with barrel recoil had been introduced. In the field artillery, light and mobile field guns with short range and limited shell effect predominated. The 15-cm heavy field howitzer and 21-cm mortar, both high trajectory weapons, more or less made up the heavy artillery together with some 42-cm mortars for the bombardment of fortifications. There were few shallow-trajectory guns.

In 1914 the flak was in the early stages of development: there were only eighteen guns, then classified as "anti-observation-balloon guns". The Imperial Navy's coastal artillery was spread amongst the fortifications on the North Sea and Baltic coasts and on the North Sea islands such as Amrun, Borkum and Heligoland. A substantial number of these guns were considered obsolete.

Comparison of German artillery weapons, 1914–1918		
	1914	1918
Field artillery	**1,069 batteries**	**2,800 batteries**
7.7-cm field guns and 10.5-cm field howitzers	6,326	11,204
Flak	18	2,576
Mountain guns	0	100
Light infantry guns	0	200
Anti-tank guns	0	1,150
Heavy artillery	**201 batteries**	**1,160 batteries**
15-cm heavy field howitzers and 21-cm mortars	2,300	7,600
13-cm guns	16	120
15-cm guns	0	232
Railway guns	0	120
42-cm mortar	7	20

Summer 1914. A 7.7-cm 96 n/a light field gun in firing position. The gunners have dug in the chassis wheels in order to obtain a range of over 5500 metres.

By the end of the Great War in November 1918 the artillery of the Army in the field amounted to 2,317 batteries with almost 24,000 guns.

Between 1914 and 1918, the field-artillery and Fussartillerie fired 290,760,000 rounds, the financial cost of this being 16.1 billion Gold Marks. This sum would be equal today to about 80 billion Euros.

What were the causes of this enormous increase in artillery? In the first weeks of the war the estimates of the military leaders as to the intensity, the geographical extension and involvement in time of a battle were found to be untenable. The losses of men and material reached a previously unknown degree. On the Western Front the exhausted enemy resorted to trench warfare. The simple infantry trench soon became a deeply layered system of defence, a true labyrinth of several lines of trenches, forward and communications trenches, underground command posts, observation-MG-and trench-mortar positions. MG firepower dominated the battlefield and was decisive for the steadfastness of the defence.

The artillery was above all the only weapons system in a position to knock out the enemy defences. Accordingly, artillery ammunition was brought up in greatly increased quantities. During the Battle of Verdun (February to December 1916) this development reached a highpoint. The field of battle, with a surface area up to 30 kms wide and 20 kms deep, was regularly ploughed over by a total of 21 million German and 15 million French shells and transformed into a lunar landscape. Only deep infantry trenches, underground command posts and galleries offered protection against artillery. Thus the number of men in the most advanced frontline could be kept small. Compared with the losses in the opening months of the war and measured by the quantities of munitions used, the casualties remained low. Instead a restructuring in depth came about.

Pre-war manouevres: this battery with four 15-cm heavy field howitzers 02 is drawn up in open firing order.

Ein Gruss aus unsern 42 cm. Geschützen.

Wenn unsre Brummer los erst gehn, Vergeht dem Feinde Hörn und Sehn.

At the beginning of the First World War the largest calibre high-trajectory artillery of the German Army made an important contribution to the penetration of the Franco-Belgian string of fortifications. This unleashed in Germany a propaganda campaign of unparalleled intensity in Germany. "Big Bertha" was celebrated as a "miracle weapon". The title of the cartoon is "A greeting from our 42-cm guns". The little rhyme reads: "When we send over our bumblebee, The enemy can neither hear it nor see."

In the rearward sectors of the front, troops and materials waited in readiness to counter-attack. Daily, especially at night, there occurred a lively traffic between the front and the rear areas. Fresh troops came forward as the relief; food, munitions and weapons were brought up. Transports were the target of long-range low-trajectory guns, only available in small numbers at the beginning of the war. Army High Command had given priority to high-trajectory guns.

Every active Army Corps in the field was given a heavy artillery batallion to make it superior to the Army Corps it faced.

The effect of the heavy 15-cm field howitzers and 21-cm mortars was impressive, they were able to lead the indirect fire advantageous in trench warfare, but their range was insufficient to meet new requirements to the extent required.

Another factor had fatal consequences: the field guns were of light construction for mobility, advantageous so long as the war remained one of mobility. By the end of 1914 that was no longer the case. The range and shell effect of the 7.7-cm light field gun 96 n/A was no longer adequate. Even the range of the 10.5-cm light field howitzer 98/09 was too short, though the shell effect was convincing.

The effort to increase range was a kind of competition which occupied the attentions of all the nations involved in the war. In Germany four different routes were followed:

(1) Modernisation of guns already in service retaining their shells, cartridge cases and powder charges
In June 1915 the 7.7-cm field gun on a howitzer chassis (barrel elevation up to +40°) arrived at the front, in the spring of 1916 the 7.7-cm field gun 16 followed with a longer barrel. The same was done for

A German field artillery position totally destroyed and abandoned. 1915/1916.

A 10.5-cm light field howitzer 16 destroyed as a result of a shell exploding in the barrel. In June 1917 the field artillery lost a total of 342 guns, one year later it was 255 (field guns and howitzers).

the 10-cm gun, the 10.5-cm light field howitzer, the 15-cm heavy field howitzer, the 21-cm mortar and the 42-cm mortar. The latter received a 30.5-cm L/30 barrel.

(2) Use of naval guns

The shortage of heavy long-range guns (heavy low-trajectory fire) was overcome by naval fast-loading guns on wheeled chassis. Krupp alone supplied 150 heavy 15-cm guns. In 1916 railway and railway roadbed guns followed. The largest calibre was the 38-cm L/45 gun.

(3) New designs of heavy field guns

While the 18-cm experimental howitzer failed to make its mark, the firms of Krupp and Rheinmetall succeeded with their 1913 design for the heavy low-trajectory 15-cm which was adopted officially in 1916. Krupp's 15-cm gun was the first to have been designed from the outset to be towed by a motor vehicle.

(4) Changes to munitions

The basic effort to increase range and shell effectiveness by a further development of the ammunition met many problems. The expense involved in the design and manufacture of an *Einheitsgeschoss* (see Glossary) led to the idea being abandoned. The value of shrapnel balls, which had been proved in the war of mobility, was disputed. For trench warfare explosive shells were preferred, the explosive component being increased.

The expenditure of ammunition increased by leaps and bounds, bringing the war economy to its limits of capacity. First came a changeover to grey cast iron shells with thicker walls. They had less room for the explosive, created fewer spinters and an increased number detonated in the gun barrel. Not until manufacture was transferred to mass production in pressed steel was a viable solution found. Fuze production also provided problems in materials and technical finish. Extraordinary efforts were required in order to meet the massive demand for propellant powder and explosive.

A completely new way to increase shell effect was provided by the introduction of gas munitions. As the war progressed, shells with longer range arrived at the front. These were the C-shells of the field artillery with a slim tip and tapered shape (see sketch and details, Glossary). The heavy and heaviest artillery shells "Haubengranaten" had a hollow cap fitted above the detonator for reduced air resistance and longer range. To illuminate the battlefield by night special starshells were used. Kartätsche (see Glossary) were reintroduced for short range defence effective up to 600 metres. After tanks had appeared

Increase in range (examples)				
Gun	Shell	Range	Shell	Range
7.7-cm field gun 16	HE	9,100 m	C-Shell	10,700 m
10.5-cm light field howitzer 16	HE	8,400 m	C-Shell	9,700 m
10-cm gun	HE	11,400 m	Hauben-shell	13,100 m
Long 15-cm heavy field howitzer 13	HE	8,500 m	Hauben-shell	8,800 m

(Left above) Exhibition of captured weapons, autumn 1914 – here 7.62-cm Putilov field guns. These Russian weapons were more efficient than the German equivalent and played an important role in equipping German units.

(Left lower) Re-activated for trench warfare – the long 15-cm gun in a prepared firing position. Photo taken in 1915.

for the first time on the Somme in the autumn of 1916, the field artillery received armour-piercing rounds 15 with an armoured head.

Infantry firepower, which had been greatly increased as the war went on (e.g. with 7.6-cm light mortars) received more light artillery. This included 7.7-cm field guns 96 n/A from artillery units. Their purpose was to take out nests of resistance by direct fire at close range. By the war's end this had resulted in special batteries of infantry guns. Amongst these "special guns" must also be counted anti-tank guns. At first field artillery and flak guns had been used to engage tanks. Because of the shortage created for their proper roles, industry was asked to design purpose-built anti-tank guns of which 600 had arrived at the front by the war's end to be operated by light-mortar batallions.

Certain parallels can be drawn in the development of flak guns. In 1906 the War Ministry had asked the Artillery Testing Commission to ascertain which field-artillery and Fussartillerie heavy guns were best suited to engage aerial targets. The first comprehensive trials showed that special guns with barrel recoil on a central pivoting chassis were best for the purpose. By the outbreak of war not many had been manufactured, and by its end the total was only 2,576, of which some were conversions of vintage or captured guns.

At the outbreak of the Great War, the German armed forces – excepting the protection forces in the colonies – had no mountain guns. During the war mountain-gun batteries and batallions were formed and by 1918 there were around 100 mountain guns in existence.

Small mobile guns of small calibre, high muzzle velocity and powerful impact were needed for anti-tank work. The photograph shows a 3.7-cm gun 1918 with the Light-mortar Reserve Batallion at Emmendingen.

10.5-cm light field howitzers 16 being accepted by Army Administration. By the war's end too many field guns had been produced and many never came to the front. They lacked trained crews, horse teams, vehicles and equipment.

The demand for special guns over-burdened the capacity of industry. The monthly production of guns for the field-artillery and Fussartillerie rose from 120 in the summer of 1914 to about 2,900 in the summer of 1918. Artillery dominated the battlefield. Its employment required not only gigantic stocks of ammunition, but firing it also resulted in rapid wear and tear to the gun barrel. By way of example: in July and August 1916 in the West, around 1,600 field and 760 heavy guns had to be replaced. Whereas enemy fire accounted for between 1.5 and 15 per cent of losses, between 85 and 98.5 per cent of losses were caused by intensive firing.

The number of shells detonating in the barrel was very high. Between mid-February and the end of May 1916 in 5.Army sector, 57.3 per cent of the 637 field guns in active use were lost to this cause, 26 per cent of the 880 heavy guns. The life of the gun barrel was

limited and varied between 2,000 rounds from the 15-cm gun 16; and 12,000 rounds from the 15-cm heavy howitzer. The limits were quickly reached. In many heavy field-howitzer batteries all barrels would have had to be changed within an eight week period. Occasionally, replacement guns would not be available resulting in captured guns arriving instead. Of the 5,074 heavy guns at the beginning of 1916, 600 of these had come from captured stocks.

Other factors to improve the effectiveness of German artillery fell within the ambit of the areas General Organisation, Composition, Training and Shooting Techniques.

1. Composition

By October 1914 ten new reserve field artillery regiments had been formed, each of three small batallions (*Abteilungen*). A battery would now have four instead of six guns due to a shortage of field guns and howitzers. In December 1914 another sixteen field artillery regiments were formed. These had only two *Abteilungen* (of four batteries each), one with twelve 7.7-cm 96 n/A field guns and one with twelve 10.5-cm light field howitzers 98/09. For the first time the field artillery preferred howitzers. In the spring of 1918 there were once more batteries with six guns: these were field position batteries without vehicles or horses. Ninety-eight batteries each with eight captured guns were formed for the Eastern Front.

Powerful motorized vehicles able to tow a heavy gun and carriage under field conditions represented a step forward for field atillery regiments, six of them being delivered in the spring of 1918. They were used for rapid assembly of artillery strongpoints.

At the beginning of the Great War, *Fussartillerie* batallions had four batteries of heavy 15-cm field howitzers or 21-cm mortars. There were also some batteries of four 10-cm guns. The make-up of batteries changed as the war progressed. Batteries with 13-cm guns and 21-cm mortars now had only three guns each, those with the new 15-cm 16 guns only two. The heaviest steep and flat trajectory gun batteries had only a single gun.

2. Shooting Techniques

From a technical point of view, the Fussartillerie adapted itself better to the changed conditions, and even became the instructor to the field artillery. For the latter, up to 1915 mobility, rapid preparedness to fire and shooting from open positions had been demanded. This would not work in trench warfare. The practices of the Fussartillerie, developed in fortress warfare, corresponded significantly better to the new conditions. To achieve penetration into the enemy defences, initially a lengthening of the period of bombardment and a higher concentration of artillery was employed. However, firing all day long robbed the German attack of the element of surprise and transformed the region into a cratered terrain difficult to negotiate. The artillery had to be employed with a more cunning approach. From amongst the various special practices introduced by the Germans subsquently, two spring to mind:

a. The Pulkov technique, using which an effective fire could be opened without previously probing for the range. This guaranteed surprise to the attackers.
b. The Bruchmüller technique. Using precise calculations, the infantry advance (beyond splinter range) was accompanied by a simple artillery bombardment. There was also a dual bombardment using splinter and gas munitions.

Both the above techniques played a decisive role in the initial German success in the Western threatre in the spring of 1918. The effectiveness of the artillery was no longer measured simply by the number of guns and great stocks of ammunition, surprise and mobility were also important.

It remained the principal task of the artillery to support the infantry whether in attack or defence. In order to fulfill this requirement, in the course of the war techniques and tactics underwent great changes. An expression of this development is reflected in the size of the artillery materials which Germany was obliged to destroy after the lost war. By virtue of the Treaty of Versailles signed on 28 June 1919, 59,897 guns and barrels, 38.75 million shells, 60.4 million live fuzes, 79,500 ammunition moulds and much other besides were reduced to scrap.

Modernization of heavy artillery had been neglected within the German Army before the war. The lack of horses and the increasing weight of heavy guns forced a rethink. This photo shows a Krupp-Daimler 100 h.p. (K.D.1) motorized artillery tug with the 8.3 tonne carriage for the barrel of the 15-cm 16 cannon. (Krupp).

21-cm mortar battery of a Fussartillerie batallion about to move out. The gun ready to be transported weighed 7.8 tonnes and was taken in two parts. 21-cm mortars with sprung chassis for the tug were not available until the war's end.

7.7-cm Field Gun 96 n/A

Calibre:	77 mm
Barrel length:	2,080 mm = l/27
Weight for transport:	1,910 kg
Weight, firing position:	1,020 kg
Weight of shells:	6.85 kg HE, 6.85 kg shrapnel
Traverse angle:	8°
Elevation/depression:	–12.9°/+15.2°
Muzzle velocity:	465 m/sec
Rate of fire:	12 rounds/min
Range, max.:	7,800 m shell with impact fuze

As a response to the introduction by the French of guns with barrel recoil in 1897, the firms of Krupp and Ehrhardt were contracted to build the new Field Gun 96. It was first tried out on manouevres in 1903, and in 1905 began to replace earlier gun types. In August 1914 5,096 guns were operational, by November 1918 there were still 3,744 in service. It was limbered to the six-horse 96 n/A. A field-gun battery would have six of these guns, the cavalry divisions were limited to four even in peacetime.

7.7-cm Light Field Gun 16

Calibre:	77 mm
Barrel length:	2,965 mm = L/35
Weight for transport:	2,256 kg
Weight, firing position:	1,325 kg
Weight of shells:	Shell 7.20, 6.6 kg shrapnel-shell, 5.99 kg C-shell.
Traverse:	4°
Elevation/depression:	–10°/+40°
Muzzle velocity:	420–525 m/sec shell, 602 m/sec C-shell.
Rate of fire:	15–20 rounds/min
Range:	9,100 m shell; 6,000 m shrapnel-shell: 10,700 m C-shell

The Field Gun 16 was a child of the war. It was developed during an economically complicated time in armaments production to answer the principal call for a gun with superior range. The men at the front welcomed it and all efforts to improve shell effect. The stock of 7.7-cm light field guns 16 rose to 3,020 guns by the end of the Great War in 1918. Four gun batteries were used in the field-gun batallions of field-artillery regiments.

7.62-cm Field Gun Putilov M.02

The ballistic performance of the Russian field gun at the beginning of the First World War exceeded that of all guns of this type in other armies. Large numbers were captured by the German Army in 1914/1915. Some were used to replace the 7.7-cm field gun 96 n/A in the batteries of Field Artillery Regiments 287 and 405. Rheinmetall produced the explosive and shrapnel shells. The greater proportion of captured Russian field guns were converted to flak and also used ammunition of German manufacture.

Calibre:	76.2 mm
Barrel length:	2,286 mm = L/30
Weight for transport:	1,945 kg
Weight, firing position:	1,040 kg
Weight of shell:	6.58 kg shell, 6.50 kg shrapnel shell, 6.41 kg mortar-shell.
Traverse:	5°
Elevation/depression:	-6°/+16°
Muzzle velocity:	588 m/sec
Rate of fire:	10 rounds/min
Range:	9,600 m shell, 7,500 m shrapnel-shell.

9-cm Field Gun C/73/91

The field gun C/73 was introduced to the field artillery of the German Army in 1874: the 88 mm calibre went to the field artillery regiments of the infantry divisions and the 78.5 mm calibre to the cavalry divisions. The uniformity of the calibre in the model C/73/91 resulted from the introduction of robust steel for the barrels, more effective explosive and constructive improvements to chassis and limber. The lack of guns and ammunition in the first year of the Great War forced the Germans to use obsolete field guns in replacement and reserve field artillery batallions.

Calibre:	88 mm
Barrel length:	2,100 mm = L723.8
Weight for transport:	2,515 kg
Weight, firing position:	1,210 kg
Weight of shell:	7.2 kg shell, 8.1 kg shrapnel-shell.
Traverse:	0° (this was achieved by shifting a beam on the chassis.)
Elevation/depression:	–15°/+16°
Muzzle velocity:	464 m/sec shell; 442 m/sec shrapnel shell.
Rate of fire:	10 rounds/min
Range:	6,500–7,100 m.

10.5-cm Light Field Howitzer 98/09

Calibre:	105 mm
Barrel length:	1,625 mm = L/16
Weight for transport:	2,260 kg
Weight, firing position:	1,225 kg
Weight of shell:	15.7-15.8 kg HE, 12.8 kg shrapnel
Traverse:	4°
Elevation/depression:	−10°/+40°
Muzzle velocity:	302 m/sec
Rate of fire:	4 rounds/min
Range:	6,300 m shell with impact fuze, 5,300 m shell with time fuze.

In 1902 the firms of Krupp and Rheinmetall received contracts to build the light field howitzer 98 as a gun with barrel recoil. In 1909 the Krupp prototype was approved officially for delivery to the troops. As a rule, one of the four batallions in the field-artillery brigades of the divisions was equipped with three batteries, each having six guns of this type. The reserve field-artillery regiments formed from October 1914 with three batallions each had one light field howitzer batallion. In August 1914 the Army had more than 1,230 10.5-cm light field howitzers 98/09.

10.5-cm Light Field Howitzer 16

Field howitzers grew in importance during the Great War. Their numbers in the field artillery batallions rose faster than the light field guns. At the same time an increase in range and shell effect was achieved. The result was the 10.5-cm light field howitzer 16 of which in November 1918 there were 3,004 in four-gun batteries in the light field howitzer batallions of the artillery regiments at the fronts.

Calibre:	105 mm
Barrel length:	2,310 mm = L/22
Weight for transport:	2,300 kg
Weight, firing position:	1,380 kg
Shell weight:	15.7 kg HE; 15.89 kg shrapnel; 15.7 kg C-shell.
Traverse:	4°
Elevation/depression:	−10°/+40°
Muzzle velocity:	400 m/sec HE, 427 m/sec C-shell.
Rate of fire:	6 rounds/min
Range:	8,400 m shell with impact fuze, 6,000 m shell with time fuze, 9,700 m C-shell.

Heavy Field Howitzer

Calibre:	149.7 mm
Barrel length:	1,620 mm = L/10.8
Weight for transport:	2,725 kg
Weight, firing position:	2,188.5 kg
Weight of shell:	42 kg 15-cm shell
Traverse:	0°
Elevation/depression:	0°/+65°
Muzzle velocity:	280 m/sec
Rate of fire:	2 rounds/min
Range:	6,050 m shell with impact fuze, 5,700 m shell with time fuze

After two years of trials, in May 1893, the heavy field howitzer was accepted. The guns, drawn by a team of six horses, were placed in batteries of four howitzers each in the heavy howitzer batallions of the Fussartillerie. In July 1914 there were 870 of these guns on hand. At the beginning of the Great War this model was considered obsolete but continued to be used operationally by the veteran reserve Fussartillerie batallions.

Experimental Howitzer 99

Calibre:	149.7 mm
Barrel length:	1,270 mm = L/12
Weight for transport:	3,055 kg
Weight, firing position:	2,380 kg
Weight of shell:	39-42 kg
Traverse:	0°
Elevation/depression:	0°/+42°
Muzzle velocity:	325 m/sec
Rate of fire:	one round/min
Range:	7,450 m

In 1899 the War Ministry in agreement with the Fussartillerie Inspectorate-General specified a new 15-cm field howitzer with a range of 7,000 metres. The following year Krupp presented its version of the gun with a short barrel recoil. Forty-two of the type were produced and tested by the troops under the designation "Experimental Howitzer 99". Its further development led to the 15-cm heavy field howitzer 02. At the outbreak of the Great War there were still 32 of the type to be found in the fortress artillery parks.

Heavy Field Howitzer 02

Calibre:	149.7 mm
Barrel length:	1,770 mm = L/12
Weight for transport:	2,710 kg
Weight, firing position:	2,035 kg–3,055 kg
Weight of shell:	39-42 kg
Traverse:	4°
Elevation/depression:	0°/+42°
Muzzle velocity:	325 m/sec
Rate of fire:	2 rounds/min
Range, maximum:	7,450 m

As a result of unfavourable trials with the experimental howitzer 99, the "lighter howitzer" was accepted into service on 18 June 1903 as the first gun in the German Army with barrel recoil. It was used in four-gun batteries of the heavy howitzer batallions of the Fussartillerie. A six-horse team drew it together with the heavy howitzer limber. In July 1914 the heavy artillery of the Field Army had over 840 guns of this type.

Bespannte Haubitzen-Battery.

Heavy Experimental Field Howitzer L/13 (Ehrhardt)

Calibre:	149.7 mm
Barrel length:	1,940 mm = L/13
Weight for transport:	2,860 kg
Weight, firing position:	2,175 kg
Weight of shell:	40.5 kg
Muzzle velocity:	370 m/sec
Rate of fire:	4 rounds/min
Range, maximum:	8,250 m

In February 1909 the War Ministry ordered from the firms of Krupp and Ehrhardt one 15-cm experimental howitzer each. Ehrhardt delivered one year later. During testing, full barrel and variable barrel recoil of the gun was found to be superior to the Krupp model. However, because it was 45 kg heavier it was not adopted. The experimental battery, ordered to be set up in May 1912, was operational in Flanders during the Great War with the Deutsche Marinekorps.

Heavy Field Howitzer 13

Calibre:	149.7 mm
Barrel length:	2,090 mm = L/14
Weight for transport:	2,820 kg
Weight, firing position:	2,195 kg
Weight of shell:	40.5–42 kg
Traverse:	5°
Elevation/depression:	0°/+45°
Muzzle velocity:	365 m/sec
Rate of fire:	4 rounds/min
Range, maximum:	8,500 m

In 1907 the Artillery Testing Commission ordered a new 15-cm howitzer with a long barrel-recoil and splinter shield. After a drawn-out development, the Krupp version was adopted in November 1913. The gun was made to be towed by the six-heavy army horse limber. The heavy howitzer batallions of the Fussartillerie had four batteries, each with four field howitzer 13's. In October 1918 there was a total of 488 guns of this type with the fighting troops.

Long Heavy Field Howitzer 13

Calibre:	149.7 mm
Barrel length:	2,250 = L/17
Weight for transport:	2,870 kg
Weight, firing position:	2,250 kg
Weight of shell:	40.3-42.3 kg
Traverse:	4°
Elevation/depression:	0°/+42°
Muzzle velocity:	377-483 m/sec
Rate of fire:	4 rounds/min
Range, maximum:	8,800 m

The most important requirement by the batallions for improving the performance of the heavy field howitzer was longer range. Therefore, during the Great War the long heavy field howitzer 13 made its appearance with a longer load chamber (see Glossary, *Ladungsraum*) and longer barrel. It was used in four-howitzer batteries within the heavy howitzer batallions. Between 1916 and the war's end 1,550 guns of this type were delivered of which 1,172 were still operational at the end of hostilities.

Long Heavy Field Howitzer 13/02

During the Great War the heavy field howitzer developed into the principal gun of the German field artillery. In order to meet the urgent need at the front, the State artillery workshops were given contracts to reinforce the chassis of the obsolete field howitzer 02s to take the more efficient barrel of the long heavy field howitzer 13. In July 1917 Krupp provided a prototype to fit the bill. About one thousand guns were converted. As was typical for this type of gun, they were used in four-howitzer batteries in the heavy howitzer batallions.

Calibre:	149.7 mm
Barrel length:	2,250 mm = L/17
Weight for transport:	3,000 kg
Weight, firing position:	2,210 kg
Weight of shell:	40.3–42.3 kg
Traverse:	5°
Elevation/depression:	0°/+45°
Muzzle velocity:	377 m/sec
Rate of fire:	4 rounds/min
Range, maximum:	8,800 m

15-cm Experimental Howitzer L/30 (Rheinmetall)

Calibre:	149.7 mm
Barrel length:	4,490 mm = L/30
Weight for transport:	6,500 kg
Weight, firing position:	4,4525 kg
Weight of shell:	44 kg
Muzzle velocity:	600 m/sec
Rate of fire:	2 rounds/min
Range, maximum:	14,500 m

(also known as Light Kartaune - see Glossary)

During the war, the Artillery Testing Commission asked for another increase in range for heavy howitzers. That led to a new design also known as "Light Kartaune" (see Glossary). Its main aim was to engage enemy artillery. Because of it's great weight, the gun was transported in two parts, (3,100 kg and 3,400 kg). In 1918 Rheinmetall turned out another two Light Kartaunen with which transportation and shooting tests were carried out.

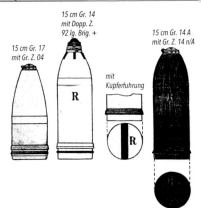

15 cm Gr. 17
mit Gr. Z. 04

15 cm Gr. 14
mit Dopp. Z.
92 lg. Brig. +

mit
Kupferfuhrung

15 cm Gr. 14 A
mit Gr. Z. 14 n/A

R

R

18.5-cm Experimental Howitzer L/22 (Krupp)

Calibre:	185 mm
Barrel length:	4,070 mm = L/22
Weight for transport:	7,210 kg
Weight, firing position:	5175 kg
Weight of shell:	80 kg
Traverse:	5°
Elevation/depression:	0°/+45°
Muzzle velocity:	440 m/sec
Rate of fire:	1 round/min
Range, maximum:	11,000 m

The 18.5-cm howitzer came into being as the result of an instruction from Army High Command which, overriding the Artillery Testing Commission, ordered from Krupp a gun of the same range as the 21-cm mortar. Three batteries each of twelve howitzers arrived at the front for trials at the beginning of 1918. The gun was dismantled into two parts (3,660 kg and 3,550 kg) for transport and hauled by twelve heavy Army horses. It was a compromise between shell effect and great range. The results at the front led to its rejection.

20-cm Howitzer M.77

(originally: heavy 8-inch cannon M.1877)

Captured Russian siege guns were a valuable addition to the stock of heavy guns held by the Fussartillerie, the reserve Fussartillerie and the local Landwehr Fussartillerie batallions of the German

Calibre:	203.2 mm
Barrel length:	3,415 mm = L/16.8
Weight for transport:	8,500 kg
Weight, firing position:	5,856 kg
Weight of shell:	78.3-90.1 kg
Traverse:	4°
Elevation/depression:	0°/+42°
Muzzle velocity:	400 m/sec
Rate of fire:	One round every three to four minutes
Range, maximum:	6,050 m (Russian ammunition), 9,600 m (German ammunition)

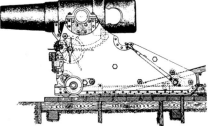

Army. Both versions of the 20-cm howitzer M.67 were used for a while. Twenty of the improved 20-cm howitzer M.77's were in German service in 1916, but only a few of these were still operational in 1918.

21-cm Mortar

(also known as 21-cm Steel Mortar)

The introduction of HE ammunition provided steep trajectory guns with a great superiority over those of flat trajectory. As early as 1893 a new 21-cm mortar had been ordered to replace the 21-cm bronze mortar. It was required to fire over a longer range in order to take advantage of the new ammunition. In July 1899 the Fussartillerie was equipped with the 21-cm steel mortars, which were grouped up in four guns per battery. There were still forty-eight in service in July 1914 being used by the non-horsedrawn reserve Fussartillerie batallions, later also by the Landwehr-Fussartillerie batallions.

Calibre:	211 mm
Barrel length:	2,110 mm = L/10
Weight for transport:	6,380 kg
Weight, firing position:	4,820 kg
Weight of shell:	83–144.5 kg
Traverse:	4°
Elevation/depression:	+6°/+70°
Muzzle velocity:	308-350 m/sec: 394 m/sec with the 83 kg heavy 21-cm shell 14
Rate of fire:	One round every three minutes
Range, maximum:	6,800–8,300 m

21-cm Experimental Mortar L/10 (Krupp)

Calibre:	211 mm
Barrel length:	2,110 mm = L/10
Weight for transport:	8,188 kg
Weight, firing position:	5,384 kg
Weight of shell:	119 kg
Traverse:	4°
Elevation/depression:	+4°/+60°
Muzzle velocity:	288 m/sec
Rate of fire:	One round per minute
Range, maximum:	7,000 m

Development work on a 21-cm mortar with barrel recoil was under way from 1899. Krupp provided a prototype for experimental purposes in 1903, with three more following up to 1906. The fourth prototype was ordered for two batteries, these arrived at the Fussartillerie Gunnery School in 1908 for testing. They were not adopted generally because the Artillery Testing Commission wanted a 21-cm mortar with 9,000 metres range.

21-cm Experimental Mortar L/12 (Rheinmetall)

Calibre:	211 mm
Barrel length:	2,530 mm = L/12
Weight for transport:	5,960 kg
Weight, firing position:	5,710 kg
Weight of shell:	120 kg
Traverse:	5°
Elevation/depression:	-5°/+65°
Muzzle velocity:	285 m/sec
Rate of fire:	One round every minute
Range, maximum:	7,000 m

In 1906 the firm Ehrhardt (later Rheinmetall) competed with Krupp in the development of a new 21-cm mortar with barrel recoil. The gun was ready in June 1908 and acquired by Army administration. Meanwhile a range of between 9,000 and 9,400 metres had been demanded. Rheinmetall provided another test model, the 21-cm experimental mortar L/15 (Rheinmetall). This was also rejected since interest was now invested in the series production of the Krupp 21-cm mortar.

21 cm Schr. 04 mit Dopp. Z. 92

21 cm Schr. 89 (abg.) mit Dopp. Z. 92

21-cm Experimental Mortar L/15 (Rheinmetall)

Calibre:	211 mm
Barrel length:	3,165 mm = L/15
Weight for transport:	8,727 kg
Weight, firing position:	6,554 kg
Weight of shell:	120–140 kg
Traverse:	4°
Elevation/depression:	0°/+65°
Muzzle velocity:	340 m/sec
Range, maximum:	9,400–10,200 m

In 1907 the Artillery Testing Commission tendered for a 21-cm mortar with a range of 9,000 metres. In 1908 Ehrhardt (later Rheinmetall) received a contract to supply such a mortar for testing purposes, and the gun was ready a year later. In the course of shooting trials it was seen to have a number of drawbacks as against the Krupp mortar. In particular, firing the mortar at the higher barrel elevations on soft ground the chassis sank in much deeper and so the gun was rejected.

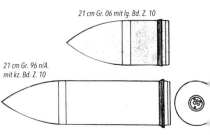

21 cm Gr. 06 mit lg. Bd. Z. 10

21 cm Gr. 96 n/A. mit kz. Bd. Z. 10

21-cm Mortar L/12

Calibre:	211 mm
Barrel length:	2,530 mm = L/11.9
Weight for transport:	11,865 kg
Weight, firing position:	7,380 kg
Weight of shell:	120 kg
Travers:	4°
Elevation/depression:	0°/+42°
Muzzle velocity:	3,367 m/sec
Rate of fire:	One round every minute
Range, maximum:	9,400 m

The 21-cm mortar was adopted officially in February 1910. In the course of 1909 all batteries of the heavy and siege artillery were equipped with the new gun. Each mortar had three vehicles drawn by six heavyweight Army horses. Each mortar batallion of the Fussartillerie had four 21-cm mortars. In July 1914, 256 of this type of gun were in service, 224 being horse-drawn and shared equally between the heavy artillery and reserve Fussartillerie, and 32 immobile in the fortifications. During the war the number of batallion mortars of the type was reduced to two batteries, each with two mortars. In October 1918 there were still 73 batteries with 219 21-cm L/12 mortars.

Long Barrel 21-cm Mortar L/14.6

Calibre:	211 mm
Barrel length:	3,063 mm = L/14.6
Weight for transport:	8,705 kg
Weight, firing position:	6,800–7,550 kg
Weight of shell:	120 kg
Traverse:	4°
Elevation/depression:	-6°/+70°
Muzzle velocity:	394 m/sec
Rate of fire:	One round per minute
Range, maximum:	10,200 m

Amongst other modifications, the requirement for an increase in the range of the 21-cm mortar led to a lengthening of the barrel. The changeover in mortar deliveries began from the end of 1916. By the end of the Great War the stock stood at 489 long-barrel 21-cm mortars spread over 163 batallions. In the summer of 1918 transportation trials ensued, the type being towed with sprung chassis behind a Krupp-Daimler K.D.1 tractor.

28-cm Howitzer L/14 i.R. (experimental howitzer Krupp)

Calibre:	283 mm
Barrel length:	3,962 mm = L/14
Weight for transport:	17,870–20,390 kg (could be partially dismantled into two or three loads for transport purposes)
Weight, firing position:	15,200 kg
Weight of shell:	285 mm
Traverse:	12°
Elevation/depression:	+20°/+65°
Muzzle velocity:	346 m/sec
Rate of fire:	One round every two minutes
Range, maximum:	9,700 m

By 1907 Krupp had developed a 20-cm howitzer for Austria. It was later produced at the company's own cost and shown to the Artillery Testing Commission. Towed by the 60-hp Daimler tug, the transportation trials were satisfactory. During the war it was used by Coastal Mortar Battery No.7.

In 1913 Krupp went on to build the demountable 28-cm experimental L/14 howitzer on a wheeled chassis. The barrel was separable, the chassis simplified for ease of transportation. At the beginning of 1915 the only example of this gun joined beta Battery No.8 on the Eastern Front.

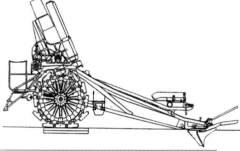

30.5-cm Mortar L/8 (*beta* unit)

In the search for an effective gun to engage fortifications which had meanwhile been reinforced by concrete stiffened with iron rods, the Artillery Testing Commission issued contracts for the development of a 30.5-cm mortar. Krupp delivered this model in 1893 and the trials continued until

Calibre:	305 mm
Barrel length:	2,440 mm = L/8
Weight for transport:	38,200 kg
Weight, firing position:	30,000 kg
Weight of shell:	333–421 kg
Traverse:	60°
Elevation:	+50°/+60°
Muzzel velocity:	310–336 m/sec
Rate of fire:	1 round every 2 minutes
Range, maximum:	8,200 m AP, 8,800 m HE

1896. The prototype proved to have a good degree of accuracy. In 1897 three batteries each with two mortars and 3,600 shells were ordered. The AP shell could penetrate an armoured cupola up to 250 mm thick. In the firing position the mortar rested on an oak bed which alone weighed 10.2 tonnes. For transportation the guns were dismantled into three parts. Two batteries were made up for road transport, the remainder for the field railway. In July 1914 there was a stock of eight heavy coastal mortars.

Heavy Coastal Mortar 09

Calibre:	305 mm
Barrel length:	4,880 mm = L/16
Weight for transport:	85,000 kg
Weight, firing position:	45,300–55,000 kg
Weight of shell:	410 kg
Traverse:	40°
Elevation:	+43°/+67°
Muzzle velocity:	395 m/sec AP, 418 m/sec HE
Rate of fire:	1 round every 2 minutes
Range, maximum:	11,900 m

Another German Army heavy gun capable of being transported by road or the field railway was the heavy coastal mortar 09. It had both barrel recoil and a pivoting chassis and required a bedplate of over 10 tons, the complicated procedure to instal it taking twelve hours. The Artillery Testing Commission ordered an experimental gun in 1908. Good trials resulted in the contract for a second mortar in November 1909. In July 1914 just before the outbreak of the First World War the two guns were assigned to *beta*-battery No.5.

30.5-cm Howitzer L/17 i.R. (*beta* unit) (Krupp experimental howitzer)

Calibre:	305 mm
Barrel length:	5,180 mm = L/17
Weight for transport:	7,100 kg
Weight, firing position:	24,500 kg
Weight of shell:	330 kg
Traverse:	10°
Elevation:	+30°/+75°
Muzzle velocity:	418 m/sec
Rate of fire:	1 round every 3 minutes
Range, maximum:	11,900 m

After the introduction of the barrel recoil on the 21-cm mortar with wheeled chassis had resulted in a noteworthy increase in performance, the Artillery Testing Commission now pressed for a similar gun of 30.5-cm calibre. In October 1912 Krupp manufactured a howitzer to this specification and presented it to the military authorities. Although the performance was satisfactory no follow-up orders ensued. The prototype was purchased from Krupp at the outbreak of war and used by Coastal Mortar Battery No.6.

Short Naval Cannon 14 L/12 (*gamma*-unit)

Calibre:	420 mm
Barrel length:	5,040 mm = L/12
Weight for transport:	93,600 kg
Weight, firing position:	42,600 kg
Weight of shell:	400–810 kg
Traverse:	10°
Elevation:	0°/+65°
Muzzle velocity:	333–500 m/sec
Rate of fire:	1 round every five minutes
Range, maximum:	9,300–12,250 m

(42-cm mortar on wheeled chassis popularly known as "Big Bertha")

This 42-cm calibre mortar later known as "Big Bertha" was designed originally as a "mine-thrower" (M-Gerät) for thin-walled shells with large explosive charge. The specifications were being continually uprated. In 1911 the contract required a 42-cm bomb-gun similar in design to 28-cm and 30-cm howitzers. The first M-Gerät was ordered in July 1911, and the second a little later. Mobility and shooting trials were both successful. For road transport the mortar was broken down into four parts, initially the towing unit was a steam locomobile. At the outbreak of the Great War there were two units available. By November 1914 the number had risen to eight. Four naval gun batteries were equipped with it. During the course of the war Krupp maunfactured a total of twelve mortars 14 L/12.

Short Naval Cannon 12 L/16 (42-cm Mortar L-16)

Calibre:	420 mm
Barrel length:	6,700 mm = L/16
Weight for transport:	76,400 kg
Weight, firing position:	42,000 kg
Weight of shell:	800-1,160 kg
Traverse:	45°
Elevation:	+43°/+66°
Muzzle velocity:	1 round every five minutes
Range, maximum:	12,500-14,100 m

(also designated *gamma*-unit)

The development of the 42-cm mortar L/16 was developed at Krupp under the strong influence of the General Staff. A gun was required capable of destroying the concrete subterranean galleries and armoured cupolas of modern fortifications. The prototype was introduced in May 1909. After the ammunition had been improved, between then and July 1914 Krupp delivered another four such mortars given the cover-name "short naval cannon 12". Repeat orders during the war resulted in the supply of ten more guns and eighteen replacement barrels.

Heavy 30.5-cm Kartaune L/30 (30.5-cm barrel L/30 on mortar chassis)

Calibre:	305 mm
Barrel length:	9150 mm = L/30
Weight for transport:	47,000 kg with bed-plate
Weight, firing position:	43,600 kg
Weight of shell:	333 kg
Traverse:	10°
Elevation:	0°/+65°
Muzzle velocity:	600 m/sec
Rate of fire:	1 round every five minutes
Range, maximum:	16,900 m

(also designated *beta*-M-unit) *Kartaune* – see Glossary

Trench warfare reduced the importance of heavy, steep-trajectory guns. The 42 tonne "M-Gerät" with its 9,300 metre range even came within range of light field guns. The Artillery Testing Commission therefore suggested placing a 30.5-cm barrel on the M-Gerät chassis. The chassis were modified and the first shooting trials with the new guns went ahead at the end of 1917. Two heavy 30.5-cm "Kartaunen" (see Glossary) arrived at the front in 1918. One gun was assigned per battery to the short naval cannon batteries No.8 and No.10.

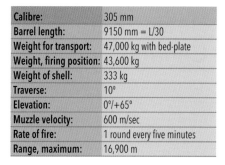

Heavy 9-cm Cannon

Large numbers of French bronze barrels had been captured in the 1870/1971 Franco-German War. They represented a good supply of of cheap material for the manufacture of German cannons.

Thus in 1879 the Fussartillerie received heavy 9-cm cannons capabale of firing field artillery ammunition, with hard bronze barrel and strengthened iron chassis. After the outbreak of war these guns and their ammunition were removed from the fortress artillery parks for use at the front, many in fixed batteries of the Landwehr- and Landsturm Fussartillerie.

Calibre:	88 mm
Barrel length:	2,110 mm = L/24
Weight, firing position:	1,308 kg
Weight of shell:	7.5 kg HE, 7.42 kg shrapnel
Traverse:	6°
Elevation/depression:	−15°/+17.5°
Muzzle velocity:	442–444 m/sec
Rate of fire:	2 rounds per minute
Range, maximum:	6,500–6,600 m

10-cm Cannon M.77

(originally 42-siege gun of the line)

This gun was licensed by Krupp Works of Essen and was manufactured at the Obuchov steel foundry. On 1 May 1900 there were more than 720 42-cannons of the line on siege chassis 1877 in the artillery parks at Russian fortresses. German troops captured a large number of these guns, together with huge stocks of ammunition, in the 1914/1915 field campaigns. 360 were put into German service, primarily with the Landwehr- and Landsturm Fussartillerie units. In the autumn of 1918 there were still 92 in existence.

Calibre:	106.7 mm
Barrel length:	3,735 mm = L/35
Weight, firing position:	2,690 kg
Weight of shell:	15.6–16.7 kg
Traverse:	0° but 37° on swivel-mounted bedplate
Elevation:	0°/+42°
Muzzle velocity:	412–518 m/sec
Rate of fire:	3 rounds every 2 minutes
Range, maximum:	8,600 m (Russian ammunition) 6,600-9,900 m (German ammunition)

10-cm Cannon with Anti-recoil Spring (Krupp)

Calibre:	105 mm
Barrel length:	3,130 mm = L/30
Weight, firing position:	2,800 kg
Weight of shell:	17.8 kg
Traverse:	4°
Elevation/depression:	–5°/+35°
Muzzle velocity:	560 m/sec
Rate of fire:	6 rounds per minute
Range, maximum:	10,300 m

At the begimnning of 1893 the Artillery Testing Commission estimated that the heavy 12-cm cannon no longer matched requirements, and a 10-cm calibre replacement was recommended. The design met some difficulties with the chassis for the Fussartillerie of the Field Army. The gun was introduced in 1899 and the same year an anti-recoil spring added (Darmarcier system). This was found unsatisfactory, and in 1901 Krupp and Ehrhardt were commissioned to build an 10-cm cannon with recoil suppression system.

10-cm Cannon 04

Krupp received from the Artillery Testing Commission a contract to build the 10-cm cannon with barrel recoil. By 1904 trials were being held using six guns in an experimental battery. In 1905 the gun was officially accepted for the Fussartillerie under the designation 10-cm cannon 04. It was a simple gun, used during the war successfully against aerial targets – aircraft and moored observation balloons. For this purpose the wheels had to be dug in or placed on a pedestal. In July 1914 there were 216 cannons of the type: in October 1918 the stock had risen to 484 guns in 121 batteries.

Calibre:	105.2 mm
Barrel length:	3,130 mm = L/30
Weight for transport:	3,509 kg
Weight, firing position:	2,807 kg
Weight of shell:	17.8–8.75 kg
Traverse:	4°
Elevation/depression:	–5°/+30°
Muzzle velocity:	560–583 m/sec
Rate of fire:	10 rounds/min
Range, maximum:	11,000–12,700 m

10-cm Cannon 14

It is noteworthy that the further development of the 10-cm cannon 04 was undertaken after experimental shooting at aerial targets in the years 1909/1910. In 1911 the Artillery Testing Commission set out the specifications and ordered prototypes from Krupp and Rheinmetall. The Krupp gun was introduced under the designation 10-cm cannon 14. The first battery was operational in May 1915 with the Fussartillerie. In the course of the war 724 guns of the type were delivered. In October 1918 there were 92 batteries with a total of 368 such guns.

Calibre:	105.2 mm
Barrel length:	3,675mm = L/35
Weight for transport:	3,400 kg
Weight, firing position:	2,814 kg
Weight of shell:	16.06–18.75 kg
Traverse:	4°
Elevation/depression:	-5°/+30°
Muzzle velocity:	560–585 m/sec
Rate of fire:	10 rounds/min
Range, maximum:	11,200–13,100 m

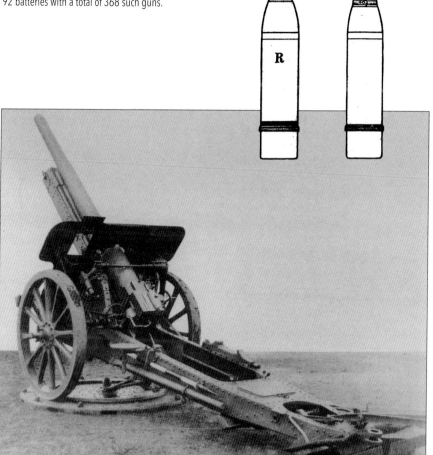

10 cm Gr. 15
mit Gr. Z. 94

10 cm Gr. 15
mit Dopp. Z. 15 (umg.)

10-cm Cannon 17

During the First World War the superiority of artillery with long range was very quickly seen. The greater gun weight was accepted. The Artillery Testing Commission wanted a longer barrel but using as much of the 10-cm cannon 14 as possible. The first 10-cm cannon 17 was delivered in the autumn of 1917. On account of the weight they were drawn in two loads (carriage and barrel cart) each by six Army heavy horses. They were used operationally by the Fussartillerie batallions, three to each battery. In October 1917 there were 90 guns of this type.

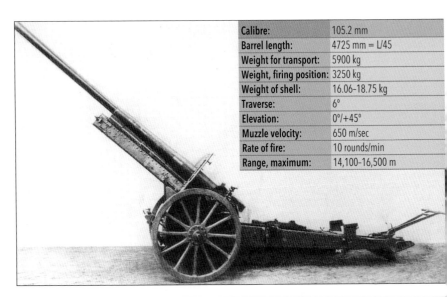

Calibre:	105.2 mm
Barrel length:	4725 mm = L/45
Weight for transport:	5900 kg
Weight, firing position:	3250 kg
Weight of shell:	16.06–18.75 kg
Traverse:	6°
Elevation:	0°/+45°
Muzzle velocity:	650 m/sec
Rate of fire:	10 rounds/min
Range, maximum:	14,100–16,500 m

10-cm Coastal Cannon L/50 on Wheeled Chassis

Calibre:	105 mm
Barrel length:	5,580 mm = L/53.14
Weight for transport:	7,140 kg
Weight, firing position:	6,500 kg
Weight of shell:	16-18.2 kg
Traverse:	6°, 160° on iron bedplate
Elevation:	0°/+42°
Muzzle velocity:	876/min
Rate of fire:	8 to 0 rounds/min
Range, maximum:	14,200-21,600 m

Four barrels of the 10-cm coastal cannon L/50 were fitted to auxiliary chassis during the Great War and delivered to the front with iron bedplates. The 1907-built guns had a gravitational block breech mechanism and a longer barrel recoil. The purpose of this makeshift solution was to co-opt surplus artillery from the Navy to augment long-range flat trajectory fire as soon as possible.

10-cm Coastal Cannon L/50 on Gunhouse Pedestal

Calibre:	105 mm
Barrel length:	5,580 mm = L/53.14
Weight, firing position:	13,555–19,055 kg
Weight of shell:	16–18.2 kg
Traverse:	360°
Muzzle velocity:	876 m/sec
Rate of fire:	10 rounds/min
Range, maximum:	14,200 m

The 10-cm coastal cannon in a gunhouse was originally used exclusively by the coastal artillery, but then operationally as a long-range flat trajectory gun on the land front. The guns were transported, similar to the 15-cm cannons on gunhouse pedestals, by the Erhardt-Bräuer load-dividing machines of the Army-Fussartillerie motor vehicle parks.

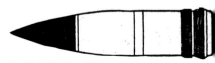

10 cm Hbschr. 16 mit Dopp. Z. 16.

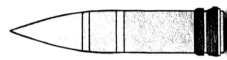

10 cm Hbgr. 16 mit Gr. Z. 04.

10.5-cm Cannon L/35 on Wheeled Chassis

(Heavy 10-cm Cannon)

Calibre:	105 mm
Barrel length:	3,670 mm = L/35
Weight for transport:	4,600 kg
Weight, firing position:	3,900 kg
Weight of shell:	17.4 kg
Traverse:	6°
Elevation:	0°/+40°
Muzzle velocity:	600 m/sec
Rate of fire:	10 rounds/min
Range, maximum:	About 12,000 m

At the instigation of the Army administration, after the outbreak of war, thirty of the 10.5-cm quick-load cannon L/35 on gunhouse pedestals were dismounted in favour of a simple wheeled chassis. This was repeated between the end of 1916 and July 1917 when the Imperial Navy's same gun type was tried out on the heavy 12-cm chassis. This makeshift solution was not very successful resulting in a reconversion to a simple wheeled chassis which could be towed by a gun carriage. In October 1918 there were four batteries with a total of sixteen guns of this type at the front.

10,5 cm Spgr. L/3,6 mit Dopp-Z. S/43

Heavy 12-cm Cannon

Calibre:	120.3 mm
Barrel length:	2,810 mm = L/23.4
Weight for transport:	4,675 kg
Weight, firing position:	2,455 kg
Weight of shell:	16.2–17 kg
Traverse:	0°
Elevation/depression:	–5°/+40°
Muzzle velocity:	398 m/sec, 439 m/sec with shell 14
Rate of fire:	2 rounds/min
Range, maximum:	7,300–7,900 m

In 1875 the Inspectorate-General of Artillery urged the withdrawal of the 12-cm cannon C/73 from the siege trains on account of its poor range. In 1880 after a long series of trials the heavy 12-cm cannon was introduced to replace it. The new gun had a hard bronze barrel with steel bore and could be transported in two sections. Six-hundred-and-twenty cannons of this type were manufactured.

At the outbreak of the First World War these now obsolete guns were used to equip fortresses. In order to make up for the shortage of guns and ammunition at the front, they were used by the Landwehr-Fussartillerie and in independent batteries.

Long 120-mm Cannon M.1878

(Originally: Canon de 120 L Modèle 1878 de Bange)

Captured French 120-mm cannons were used in the German Army by the Landwehr- and Landsturm-Fussartillerie. In 1916 there were fifteen guns of this type, and forty in the autumn of 1918. After the stock of captured 120-mm ammunition for this gun ran out, it was reproduced in Germany.

Calibre:	120 mm
Barrel length:	3,250 mm = L/27.1
Weight for transport:	3,500 kg
Weight, firing position:	2,700 kg
Weight of shell:	18-20.35 kg
Traverse:	0°
Elevation/depression:	-17°/+28°
Muzzle velocity:	613 m/sec
Rate of fire:	One round/min.
Range, maximum:	11,200-12,700 m (French ammunition), 8,500-8,800 (German ammunition)

13-cm Cannon

Calibre:	135 mm
Barrel length:	4,725 = L/35
Weight for transport:	7,847 kg
Weight, firing position:	6,000 kg
Weight of shell:	40.8–42 kg
Traverse:	4°
Elevation/depression:	–5°/+26°
Muzzle velocity:	695 m/sec
Rate of fire:	3 rounds/min
Range, maximum:	14,500–16,500 m

Development of the 13-cm cannon dates back to 1901 when Krupp received the contract to manufacture a gun to take out armoured underground command posts in fortress installations. Its variable history began in 1909 when acceptance was authorized but production halted prematurely. In July 1914 there were only sixteen 13-cm cannons in existence.

The planned completion of another sixteen cannons in the case of mobilization was interrupted in favour of the increased supply of steep-trajectory guns. Not until the troops at the front insisted on more heavy flat-trajectory guns was production resumed. Because heavy towing machines were always required for the transport of heavy cannons, sprung chassis and barrel carts were introduced. The sprung 13-cm cannon was first delivered in October 1917 and was used in batteries of three in the mixed Fussartillerie and reserve Fussartillerie batallions. Krupp delivered a total of 158 of this type, of which in Ocotber 1918 120 units remained at the front in forty batteries.

Long 155-mm Cannon M.1877

(originally Canon de 155 L Modèle 1877 de Bange)

This was an old French fortress and siege gun. In 1916 the Landwehr- and Landsturm-Fussartillerie still had twenty-eight of them. Of these, twenty-four remained in service by autumn 1918.

Calibre:	155 mm
Barrel length:	4,200 mm = L/27.1
Weight for transport:	6,622 kg
Weight, firing position:	5,800 kg
Weight of shell:	40.59–43.2 kg
Traverse:	4°
Elevation/depression:	-10°/+28°
Muzzle velocity:	223–534 m/sec
Rate of fire:	1.5 rounds/min
Range, maximum:	11,500 m

15-cm Ring Cannon

In 1871 trials began for a new 15-cm cannon. The problem areas were the durability of bronze and cast steel barrels. Krupp suggested a cast steel jacket which proved satisfactory. After the ammunition and detonator had also been improved, the gun was introduced in 1872 as a 15-cm jacket cannon with an iron chassis. During the First World War hydraulic braking was fitted in the bedplate. This type was used operationally in batteries of four guns each by the Landwehr- and Landsturm-Fussartillerie batallions.

Calibre:	149.1 mm
Barrel length:	3,440 mm = L/23
Weight for transport:	5,803 kg; 7,280 kg transported in two parts
Weight, firing position:	4930 kg
Weight of shell:	42 kg shell 14, 35.5 kg shrapnel shell
Traverse:	0°
Elevation/depression:	-5°/+37°
Muzzle velocity:	396 m/sec
Rate of fire:	2 rounds/min
Range, maximum:	7,900–8,050 m

Long 15-cm Cannon

Calibre:	149.1 mm
Barrel length:	4,470 mm = L/30
Weight for transport:	6,922 kg, 8,277 kg transported in two parts
Weight, firing position:	6,032 kg
Weight of shell:	44 kg shell 14 with cap: 35.5 kg shrapnel shell
Traverse:	0°
Elevation/depression:	-4°/+40°
Muzzle velocity:	483 m/sec
Rate of fire:	2 rounds/min
Range, maximum:	9,500–12,000 m.

The performance of the 15-cm ring cannon was bettered by foreign developments in the 1880s, causing the Artillery Testing Commission to request new designs. Testing of these began in 1885. The best results came from the Krupp steel jacket-ring barrels. Development work terminated in 1889 but adoption of the gun was delayed because new propellant powder and explosives (shell filling 88 – see Glossary) made changes in chassis design necessary. During the First World War, long 15-cm cannons remained in service in four-gun batteries at the front in Landwehr and Landsturm Fussartillerie batallions.

Heavy 15-cm Cannon L/30 on Wheeled Chassis

Calibre:	149.1 mm
Barrel length:	4,470 mm = L/30
Weight for transport:	10 tonnes
Weight, firing position:	10,050 kg with wheel tracks
Weight of shell:	40 kg
Traverse:	10°
Elvation:	maximum + 35°
Muzzle velocity:	609 m/sec
Range, maximum:	12,100 m

This type from 1883 had no barrel recoil and fired from a makeshift platform. After a round was discharged, stop-wedges returned the chassis to the correct firing position. It was obsolete and belonged to the 150 heavy 15-cm guns in naval stocks which had been modified to wheeled chassis for use on land and coastal fronts. They were used operationally by the Fussartillerie in batteries of two guns on wheeled chassis.

15 cm Sprgr. L/5
(Haube)
mit Gr. Z. 16

15-cm Cannon on Gunhouse Chassis

Calibre:	149.3 mm
Barrel length:	5,960 mm = L/40
Weight for transport:	22,500 kg
Weight, firing position:	12,120 kg
Weight of shell:	51.55-52.5 kg
Traverse:	120°
Elevation/depression:	-4°/+30
Muzzle velocity:	688 m/sec
Rate of fire:	1 round/min
Range, maximum:	19,500 m

This type of gun had been developed for the fortresses at Metz and Strasbourg, at that time in German Alsace-Lorraine. Despite their great weight the guns could be repositioned inside the fortresses by road or rail under tow. During the First World War they emerged from the two fortresses and were moved around the Army-Fussartillerie motor parks on low-loaders. Steam locomobiles were also used. The gun was bedded for firing, a procedure which would take two to three hours before the first round was fired. In July 1914 there were twelve 15-cm cannons of this type in service, although initially only two were suitable for road transport.

15-cm Fast-load Cannon L/40 on Wheeled Lafette

(Heavy 15-cm Cannon)

Before the First World War, Naval Command had contacted Krupp regarding land uses for their surplus heavy guns. The barrels were fitted on makeshift wheeled chassis as a trial, but the short barrel recoil

made them heavy and unmanouevrable. In 1915 most were made available to the Army where they were used operationally in two-gun heavy-cannon batteries of the Fussartillerie and reserve Fussartillerie regiments. In October 1918 there were still twenty-six batteries with fifty-two 15-cm fast-load cannons L/40 and L/30 on wheeled chassis. A gun would be separated into two parts for transport, often requiring a team of up to fourteen horses to haul the burden. Later, motorized tractors was used for this purpose.

Calibre:	149.1 mm
Barrel length:	5960 mm = L/40
Weight for transport:	21,150 kg
Weight, firing position:	18,950 kg
Weight of shell:	44 kg
Elevation/depression:	-8°/+32°
Muzzle velocity:	750 m/sec
Rate of fire:	1–2 rounds/min
Range, maximum:	18,700 m

15-cm Fast-load Cannon L/45

(Central pivot chassis on river barge)

Calibre:	149.1 mm
Barrel length:	6,700 mm = L/45
Weight, firing position:	55.5 tonnes
Weight of shell:	44 kg
Traverse:	180°
Muzzle velocity:	840 m/sec
Rate of fire:	1 round every 3 mins
Range, maximum:	22,650 m

This type of gun had been installed between 1908 and 1912 as medium artillery on the Imperial Navy's pre-Dreadnought battleships. Some of them were unshipped and transferred to river barges to provide mobile firing positions on the French canals. In October 1918 there were two batteries at the front, each with one gun of this type on a barge chassis.

15 cm Sprgr. L/5 (Haube) mit Gr. Z. 16.

15-cm Fast-load Cannon L/45

(Cental pivot chassis on railway low-loader frame "Nathan")

Calibre:	149.1 mm
Barrel length:	6,700 mm = L/45
Weight, firing position:	55.5 tonnes
Weight of shell:	44 kg
Traverse:	50° (180° with supports)
Muzzle velocity:	840 m/sec
Rate of fire:	1 round every 3 minutes
Range, maximum:	22,650 m

Another use for this type of gun, originally installed between 1908 and 1912 as medium artillery on the Imperial Navy's pre-Dreadnought battleships, was as a railway gun with supports enabling it to traverse through 180° on a low-loader waggon. In October 1918 there were ten "Nathans" in five batteries providing heavy low-trajectory fire at the front.

15 cm Sprgr. L/5 (Haube) mit Gr. Z. 16.

15-cm Cannon 16 (Rheinmetall)

Calibre:	149.3 mm
Barrel length:	6,410 mm = L/43
Weight for transport:	11,950 kg
Weight, firing position:	9,100 kg, 9,240 kg with wheel track
Weight of shell:	51.55–52.5 kg
Traverse:	6°
Elevation/depression:	–5°/+40°
Muzzle velocity:	749 m/sec
Rate of fire:	1 round per minute
Range, maximum:	20,200–22,800 m

Experiments with 15-cm cannons with barrel recoil suppression were already under way by 1907 but dicontinued in favour of a 13-cm cannon, and not resumed until the spring of 1913 when the Fussartillerie requested 15-cm cannons with long range for the border fortifications. Not until September 1914 did Krupp and Ehrhardt (later Rheinmetall) receive contracts for experimental guns of this type. They were ready by 1915 and underwent trials with great haste. Nevertheless, the first Rheinmetall guns were not delivered until the spring of 1917. They were transported in two parts, each drawn by six heavy Army horses, later by powered machines. The Rheinmetall guns of this type went to the two-gun batteries of the Fussartillerie batallions. By October 1918 there were twenty-two guns of this type at the front.

15-cm Cannon 16 (Krupp)

Calibre:	149.3 mm
Barrel length:	6,410 mm = L/43
Weight for transport:	14,405 kg
Weight, firing position:	10,140 kg
Weight of shell:	40.8–42 kg
Traverse:	84°
Elevation/depression:	-3°/+42°
Muzzle velocity:	749 m/sec
Rate of fire:	1 round per minute
Range, maximum:	20,200–22,800 m

Work was already in progress in 1907 on 15-cm cannons with barrel recoil for the Fussartillerie but was suspended on the orders of the War Ministry, which preferred the light 13-cm cannon. Subsequently, Krupp exported 15-cm cannons with barrel recoil on wheeled chassis and was involved in 1913/1914 in the development of 15-cm cannons for the Fussartillerie of the German Army. An experimental gun was delivered in 1915. The official introduction was authorized in August 1916 although deliveries had already begun in the spring of that year.

The 15-cm cannon was transported in two sections – chassis and barrel-carriage – by Krupp-Daimler KD1 towing machines.

The mixed power-traction batallions of the Fussartillerie had two 15-cm cannons 16 per battery. Pedigree-equipped batallions had batteries of three 15-cm cannons. In October 1918 there were 212 Krupp guns of this type at the front.

17-cm Fast-load Cannon L/40 on Wheeled Chassis

This gun was also a stopgap from the Imperial Navy. It was much heavier than comparable 15-cm guns and was transported in three sections. The first of them were received at the front by the Fussartillerie non-horse drawn batteries in March 1917, each battery had two guns. In October 1918 the German Army had over five of these batteries with a total of ten guns of this type.

Calibre:	172.6 mm
Barrel length:	6,900 mm = L/40
Weight for transport:	45,010 kg
Weight, firing position:	23,500 kg
Weight of shell:	63 kg
Traverse:	8° (360° on iron bedplate)
Elevation:	Maximum +47.5°
Muzzle velocity:	815 m/sec
Rate of fire:	One round per minute
Range, maximum:	24 kms

17-cm Fast-load Cannon L/40 "Samuel"

Calibre:	172.6 mm
Barrel length:	6,900 mm = L/40
Weight for transport:	61,500 kg
Weight, firing position:	61,500 kg
Weight of shell:	62.8 kg
Traverse:	10°
Elevation:	Maximum +47.5°
Muzzle velocity:	815 m/sec
Rate of fire:	One round per second
Range, maximum:	24,020 m

In 1916 a number of the fast-load cannons on wheeled chassis taken from naval stock were converted swiftly into railway cannons. In this way

they could be withdrawn more quickly from exposure to long-range enemy flat-trajectory guns. There were sixteen examples of the "Samuel" by October 1918, operational at the front in two-gun batteries.

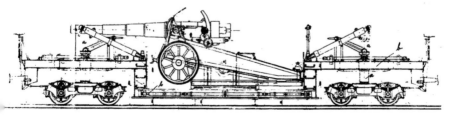

21-cm Fast-load Cannon L/40 *Peter Adalbert*

Calibre:	209.3 mm
Barrel length:	8,400 mm = L/45
Weight for transport:	110,500 kg
Weight, firing position:	110,500 kg
Weight of shell:	29 kg
Traverse:	60°
Elevation:	Maximum +45°
Muzzle velocity:	770 m/sec
Rate of fire:	Three rounds every two minutes
Range, maximum:	17,500–25,580 m

The 12,300 kg heavy barrels for the L/40 were taken from the 1901/1902-built large cruisers of the Imperial Navy and bedded on railway firing platforms in 1916. The same was done for the L/45 version. Both of these railway guns are known as "Peter Adalbert". In October 1918 there remained at the front one heavy flat-trajectory battery with a single L/40.

21-cm Fast-load Cannon L/45 *Peter Adalbert*

Calibre:	209.3 mm
Barrel length:	9,450 mm = L/45
Weight for transport:	107 tonnes
Weight, firing position:	107 tonnes
Weight of shell:	115 kg
Traverse:	60°
Elevation:	0°/+45°
Muzzle velocity:	800 m/sec
Rate of fire:	Three rounds every two minutes
Range, maximum:	18,700–26,400 m

In 1916 more efficient 21-cm ex-Imperial Navy fast-load cannons L/45 were installed on firing platforms. Also known as "Peter Adalbert", the L/45 were used operationally in single-gun batteries to provide the heaviest flat-trajectory fire the Army had to offer. By October 1918 all L/45 "Peter Adalbert" guns had been lost.

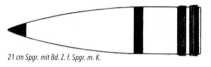

21 cm Spgr. mit Bd. Z. f. Spgr. m. K.

21 cm Spgr.
L/4,9 (Haube) mit Bd. Z. f. Spgr. m. K.

24-cm Fast-load Cannon L/30 *Theodor Otto*

Calibre:	238 mm
Barrel length:	7,200 mm = L/30
Weight for transport:	113 tonnes
Weight, firing position:	113 tonnes
Weight of shell:	148.6 kg
Traverse:	180° on bedplate
Elevation:	Maximum +45°
Muzzle velocity:	640 m/sec
Rate of fire:	Two rounds/min
Range, maximum:	18,700 m

(Bedded on rail waggon chassis)

The gun "Theodor Otto" was an old model operational from a 1917 firing platform on a railway chassis. The batteries of the heaviest flat-trajectory fire were each given one gun of this type. In 1918 none were to be found at the front.

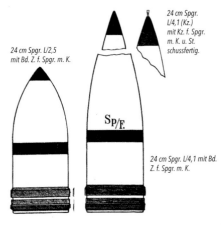

24 cm Spgr. L/2,5 mit Bd. Z. f. Spgr. m. K.

24 cm Spgr. L/4,1 (Kz.) mit Kz. f. Spgr. m. K. u. St. schussfertig.

24 cm Spgr. L/4,1 mit Bd. Z. f. Spgr. m. K.

24-cm Fast-load Cannon L/40 *Theodor Karl*

(Bedded on railway chassis)

Calibre:	238 mm
Barrel length:	9,600 mm = L/40
Weight for transport:	109 tonnes
Weight, firing position:	109 tonnes
Weight of shell:	151–190 kg
Traverse:	180°
Elevation:	Maximum, +45°
Muzzle velocity:	810–840 m/sec
Rate of fire:	Two rounds/min
Range, maximum:	23,100–26,600 m

This type of gun had been installed originally aboard the pre-Dreadnought battleships *Mecklenburg*, *Schwaben* and *Wettin*. The Imperial Navy made twelve barrels available as railway guns. They were mounted on a platform in such a way that they could be used on a rail chassis, turntable chassis or both combined. The L/40 "Theodor Karl" guns belonged in the group of heaviest flat-trajectory guns which the German Army possessed and were used in single-gun batteries to engage especially important targets. In October 1918 there were twelve such batteries at the front.

28-cm Fast-load Cannon L/40 *Bruno*

Calibre:	230 mm
Barrel length:	11,200 mm = L/40
Weight for transport:	156 tonnes
Weight, firing position:	156 tonnes
Weight of shell:	284 kg
Traverse:	8° (180°–360° on bedplate)

The L/40 "Bruno" type guns bedded on a railway chassis were 28-cm C/1901 main guns from the 1902/1904-built Braunschweig-Class pre-Dreadnought battleships. In 1917, mounted on firing platforms, they formed part of the heaviest flat-trajectory groups in single-gun batteries. In October 1918 there was still one gun of this type at the front.

38-cm Fast-load Cannon L/45 *Max*

Calibre:	380 mm
Barrel length:	17,100 = L/45
Weight for transport:	347,340 kg
Weight, firing position:	240 tonnes without turntable
Weight of shell:	400-750 kg
Traverse:	2°, 360° on turntable
Elevation:	0°/+18° without turntable:
	0°/+55° with turntable
Muzzle velocity:	890-1,041 m/sec
Rate of fire:	One round/min
Range, maximum:	24 kms-47.5 kms

This L/45 gun had been developed by Krupp for the new battleships of the Bayern- and Baden-Classes. During the First World War the completion of these ships had been held back in favour of U-boat production. The guns were therefore released for use by the naval coastal artillery in Flanders and for the heaviest flat-trajectory fire. Eventually only one 38-cm barrel L/45 was fitted to a platform so as to be capable of firing from a turntable or railway chassis. For the turntable installation a pit of 12 metres diameter and 4 metres deep had to be excavated. In October 1918 a single L/45 "Max" battery (the Saxon 38-cm battery 1050) remained at the front.

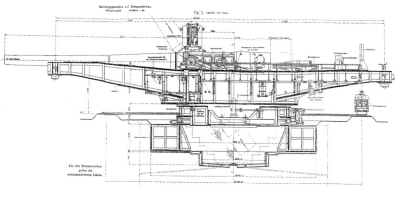

21-cm Cannon L/162

(on railway chassis) The "Paris" or "Wilhelm" Gun

Calibre:	209.3-232 mm (barrels were frequently rebored)
Barrel length:	33,910 mm = L/162
Weight, firing position:	1,490 tonnes
Weight of shell:	120 kg
Traverse:	60ª
Elevation:	0°/+55°
Muzzle velocity:	1,425–1,646 m/sec
Rate of fire:	One round every two minutes
Range, maximum:	128 kms, (summit of trajectory 40 kms)

In the autumn of 1914 the Supreme Command of the Army wanted the Imperial Navy to bombard Dover from the Belgian coast. The Reich Navy Office turned to Krupp where work was proceeding at a feverish pace on long-range cannons and ammunition. In July 1917 the first shooting trials were held at the new Altenwalde range near Cuxhaven, other trials were made at Meppen. Initially, the very long 21-cm barrel was fitted as an experiment inside the barrel jacket of the 38-cm L/45 cannon. Eventually, the railway firing platform chassis for the 21-cm cannon was developed from that designed for the 38-cm gun.

In order for the 120 kg shells to reach the stratosphere, a muzzle velocity of 1,600 metres per second was required. A multi-stage propellant with 200 kg of powder was needed, for which the loading chamber of the shell was manufactured almost 4 metres in length.

Three 21-cm L/162 cannons were completed. The seven barrels manufactured had to be rebored in March 1918 for wear and tear. The guns were used operationally in the spring of 1918 to fire 367 shells at the city of Paris from distances of between 87 and 128 kilometres.

2-cm Aircraft Gun Becker

Calibre:	20.1 mm
Barrel length:	804 mm = L/40
Weight, firing position:	33.6 kg (barrel and breech only)
Weight of shell:	0.140 kg
Traverse:	360°
Elevation/depression:	–5°/+80°
Muzzle velocity:	500 m/sec
Rate of fire:	120 rounds/min
Range, maximum:	3,000 m
Shell, maximum altitude:	2,000 m

The Becker 2-cm anti-aircraft gun was developed at the end of 1917 at the prompting of the War Ministry following the tank battle at Cambrai. It appeared at the front mounted provisionally as an anti-tank gun. In May 1918, on the orders of Army Supreme Command, it was withdrawn and 131 Becker cannons were modified for the anti-aircraft role. They were fitted with a Koebe swivelling base and a front ring sight for use as provisional flak principally against Army aircraft (bombers). The first fifty guns were scheduled to be ready in December 1918, mass production was to follow with effect from January 1919.

3.7-cm Revolver Cannon (as Flak)

Calibre:	37 mm
Barrel length:	1,190 mm = L32.2
Weight, firing position:	211 kg (with breech mechanism)
Weight of shell:	0.465 kg
Traverse:	5°
Elevation:	0°/+80°
Muzzle velocity:	540 m/sec
Rate of fire:	40 rounds/min
Range, maximum:	6,000 m
Shell, maximum altitude:	2,200 m

The 3.7-cm revolver cannons C73/88 were used until 1893 as flanking weapons in fortifications on account of the *Kartätsch* or shrapnel effect of their munition (see Glossary). Some found their way into the trenches at the beginning of the First World War. In January 1915, 180 of them were in use as trench guns while another portion of the total served as makeshift flak. The gun consisted of five steel tubes, baseplate and mechanism with hand crank: various designs of mount existed. As a rule the gun was rotatable on field-cart wheels either fixed or on mobile installation (an adapted field cart) for operations.

3.7-cm Maxim Cannon on Wheeled Chassis

(on mobile mount – Krupp)

Calibre:	37 mm
Barrel length:	1,105 mm = L/30
Weight for transport:	2,020–2,130 kg
Weight, firing position:	790–1,060 kg
Weight of shell:	0.465 kg
Traverse:	360º
Elevation/depression:	–3º/+80º
Muzzle velocity:	540 m/sec
Rate of fire:	250 rounds/min
Range, maximum:	6,000 m
Shell, maximum altitude:	2,500 m

In 1915 the Reich Navy Office made available the 3.7-cm Maxim machine cannon originally used aboard torpedo-boats. The initial manufacturer was the Deutsche Waffen und Munitions factory in Berlin. Krupp modified the guns to meet the specification for flak. There were two designs, one on the heavy chassis of the Fussartillerie for the 9-cm field-gun C/73/91 and another on a fixed pivot. The first hundred of these water-cooled flak guns were ready by September 1915. They were used for air defence both at home and at the front, in the latter capacity to defend German tethered observation balloons.

3.7-cm Pedestal Flak L/14.5 (Krupp)

Calibre:	37 mm
Barrel length:	971.5 mm = L/14.5
Weight for transport:	670 kg (with 176 rounds)
Weight, firing position:	215 kg
Weight of shell:	0.45 kg
Traverse:	360°
Elevation/depression:	–5°/+80°
Muzzle velocity:	350 m/sec
Rate of fire:	120 rounds/min
Range, maximum:	5,000 m
Shell, maximum altitude:	2,200 m

These flak guns were also intended originally for another purpose. Krupp had developed the gun as an armament for airships. As flak, the gas-pressure loader was set on a pedestal with pivoting bracket and could be dismantled into four individual parts for transport over short distances, otherwise a cart drawn by two horses was used. The first twenty guns were delivered at the end of December 1917: by the time of the Armistice in November 1918 there were 150. They were used to engage infantry low-level fighter-bomber aircraft.

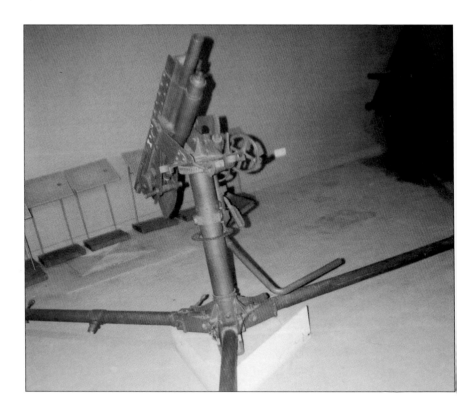

7.7-cm Flak L/27 (Krupp)

Calibre:	77 mm
Barrel length:	2,080 mm = L/27
Weight for transport:	7,030 kg
Weight, firing position:	7,030 kg (1082 kg, gun only)
Weight of shell:	6.85 kg
Traverse:	360°
Elevation/depression:	–5°/+70°
Muzzle velocity:	465 m/sec
Rate of fire:	20–25 rounds/min
Range, maximum:	7,800 m
Shell, maximum altitude:	4,250 m

(on Krupp-Daimler 14 light truck chassis)

In 1912, Krupp and Rheinmetall delivered the two types of 7.7-cm flak gun introduced in 1914 under the designation "Leichte Kraftwagengeschütze M.1914". Their great mobility was achieved by having them installed on an all-wheel drive truck chassis capable of a maximum speed of between 50 and 60 km/hr. During the course of the war, truck and gun were being continually improved. In 1918 there were 156 of these guns in Army Kraftwagen-flak batteries in almost all theatres of war.

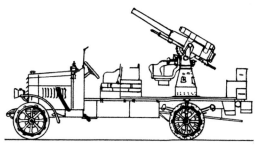

7.7-cm Field Gun 96 n/A

(as flak on provisional pedestal)

Calibre:	77 mm
Barrel length:	2,080 mm = L/27
Weight, firing position:	1,020 kg
Weight of shell:	6.85 kg
Traverse:	360°
Elevation/depression:	-5°/+52°
Muzzle velocity:	465 m/sec
Rate of fire:	10-15 rounds/min
Range, maximum:	7,800 m
Shell, maximum altitude:	4,250 m

The flak guns on hand at the outbreak of the Great War – in October 1914 there were 36 guns – were only sufficient in number for the protection of important bridges and the airship hangars. Before new flak guns came to the front, or various kinds of old or captured field guns were delivered, the troops had to come up with their own pedestal design. Particularly for the 7.7-cm field gun 96 n/A there were numerous makeshift creations. They were used for the defence of property or installations, as a rule two guns being selected from the battery for the purpose. They were obviously not employed to engage classic artillery targets within the battery or batallion.

7.7-cm Field Gun 96 n/A

(as flak on Schnetzler Pedestal)

Calibre:	77 mm
Barrel length:	2,080 mm = L/27
Weight for transport:	1,750 kg
Weight, firing position:	1,021 kg
Weight of shell:	6.85 kg
Traverse:	360°
Elevation:	Up to +75°
Muzzle velocity:	465 m/sec
Rate of fire:	10–15 rounds/min
Range, maximum:	7,800 m
Shell, maximum altitude:	4,250 m

To make a provisional flak out of the 7.7-cm field gun 96 n/A, the Schnetzler Pedestal was manufactured in the siege workshops at Longuyon. The design was used throughout the war, but required a lot of time for erection and changes of location. It remained a makeshift solution to adjust for the lack of flak guns. By the end of the Great War there were forty-four of these Schnetzler Pedestal guns operational.

7.7-cm Flak

(Makeshift solution using 7.7-cm field gun 96 n/A on howitzer chassis 98/09.)

This gun was a makeshift flak made in small numbers by fitting a 7.7-cm field gun 96 n/A on a 10.5-cm field howitzer 98/09 chassis. The latter offered the advantage of substantial elevation. A similar idea was the combination of a 7.7-cm Belgian field-gun barrel on a Russian 122-mm howitzer M.09 chassis. These flak guns were grouped in flak-platoons attached to a field-artillery regiment of an Army infantry division.

Calibre:	77 mm
Barrel length:	2080 mm = L/27
Weight of shell:	6.85 kg
Traverse:	4°, 360° on pedestal
Elevation/depression:	−10°/+70° on pedestal
Muzzle velocity:	465-510 m/sec
Rate of fire:	6 to 10 rounds/min
Range, maximum:	7,800–8,500 m
Shell, maximum altitude:	4,200–4,650 m

7.7-cm Flak L735 (Krupp)

(Adaptation of French 7.5-cm field gun M.97 L/46)

Calibre:	77 mm
Barrel length:	2,721 mm = L/35
Weight for transport:	2,050 kg
Weight, firing position:	1,250 kg
Weight of shell:	6.85 kg
Traverse:	360° (with wheel track)
Elevation/depression:	High +38°/+60°, low -1°/+21°
Muzzle velocity:	487 m/sec
Rate of fire:	10–15 rounds/min
Range, maximum:	7,200 m
Shell, maximum altitude:	c. 4,000 m

The shortage of flak at the beginning of the First World War forced the War Ministry to improvise. In negotiations with Krupp it was agreed that large numbers of captured French 7.5-cm field guns would be converted to flak weapons. The barrels were bored down to receive 7.7-cm calibre shell, the load chamber was enlarged and the chassis structure modified. In order to obtain an all-round traverse, the gun was anchored forwards and circled a wheel-track. For transport, the six-horse French or German limber (with eighteen to twenty-four shells) was used. Krupp delivered 394 flak guns of this type, Rheinmetall another 68 on a pedestal base. By the spring of 1916 every field division had a flak platoon with two horse-drawn guns attached to its field-artillery regiment.

7.62-cm Flak L/30

From 1915 in addition to French field guns, captured Russian guns were available for conversion into flak in large numbers. Here again improvisation was necessary. The guns pivoted on a pedestal and were equipped with a new sight. German 7.62-cm field-gun 96 n/A ammunition was used. The version shown here was transported on a collapsible bogie drawn by six horses. Rheinmetall delivered 120 of this type, and another 120 with a pedestal base. They were operational in flak platoons in the field-artillery regiments of the field divisions, but later only on German soil to protect important installations.

Calibre:	76.2 mm
Barrel length:	2,286 mm= L/30
Weight for transport:	2,440 kg (total 4,000 kg)
Weight, firing position:	1,350 kg
Weight of shell:	6.85 kg
Traverse:	360°
Elevation:	+6°/+70°
Muzzle velocity:	410–588 m/sec
Rate of fire:	15 rounds/min
Range, maximum:	7,400–9,400 m
Shell, maximum altitude:	4,050–5,600 m

7.62-cm Flak L/30 (Krupp) on Pedestal

Calibre:	76.2 mm
Barrel length:	2,286 mm = L/30
Weight for transport:	3,100 kg
Weight, firing position:	2,100 kg
Weight of shell:	6.85 kg
Traverse:	360°
Elevation:	0°/+70°
Muzzle velocity:	410–588 m/sec
Rate of fire:	5 rounds/min
Range, maximum:	7,400–9,400 m
Shell, maximum altitude:	4,050–5,600 m

(Adapted Russian 7.62-cm field cannon M.02)

This version of a converted Russian 7.62-cm field gun for aerial targets was finally delivered in December 1917. The barrel, rotatable through 360°, was mounted on a pedestal, the gun being stabilized by four spread outriggers which guaranteed high stability when shooting. The transition from transport to firing position, however, took some time. The guns were also used on railway waggons. In October 1918 there were 120 of this type. They were deployed in flak platoons of the field-artillery regiments and finally to defend the Homeland.

7.7-cm Flak L/35 (Krupp)

Calibre:	77 mm
Barrel length:	2,695 mm= L/35
Weight for transport:	3,125 kg
Weight, firing position:	2,500 kg
Weight of shell:	6.85 kg
Traverse:	360°
Elevation:	0°/+70°
Muzzle velocity:	510 m/sec
Rate of fire:	20-25 rounds/min
Range, maximum:	8,500 m
Shell, maximum altitude:	4,750m

In October 1915 Krupp and Rheinmetall received orders to supply thirty-two horse-drawn pedestal flak guns of increased performance. The guns were ready from December 1916 and were intended primarily for deployment on the unfavourable roads and terrain in the eastern theatre of war. Pedestal and gun was drawn by a team of six horses. The Krupp version was considerably heavier than that of Rheinmetall, and the pedestal mount had some drawbacks. Both models could be readied for action very quickly and had a fast rate of fire. They were operational in the flak platoons of field-artillery regiments.

9-cm Flak C/73 and C73/91

(On Wohlgemuth frame pedestal, as improvised flak)

Calibre:	88 mm
Barrel length:	2,100 mm = L23/9
Weight of shell:	7.42-7.71 kg
Traverse:	360°
Elevation/depression:	-1°/+70°
Muzzle velocity:	435-464 m/sec
Rate of fire:	3 to 5 rounds per min.
Range, maximum:	6,400-7,100 m
Shell, maximum height:	3,600-4,000 m

The release of the 7.7-cm field gun 96 n/A for combination into various mounting arrangements as a flak gun weakened the field artillery in ground combat. Therefore, from December 1915 recourse was had to the 9-cm field gun C/73 and C/73/91 which had been installed in the meantime as a fortress armament. The guns were bolted down permanently into wooden platforms on a pedestal framework known as the Wohlgemuth system. They were also used on other designs of bearer platform in rearward operational areas of the Field Army and in Germany for air defence. In 1918 there were still 614 of these makeshift flak guns in existence.

8.8-cm Flak L/45 (Rheinmetall)

Calibre:	88 mm
Barrel length:	3,965 mm = L/45
Weight for transport:	7,200 kg
Weight, firing position:	2,795 kg (without trailer)
Weight of shell:	9.6 kg
Traverse:	360°
Elevation/depression:	–4°/+70°
Muzzle velocity:	542–785 m/sec
Rate of fire:	10 rounds/min
Range, maximum:	9,800–10,800 m
Shell, maximum altitude:	5,200–6,850 m

(8.8-cm Flak L/45 on Motor-Towed Trailer – Rheinmetall)

As the result of discussions in April 1915 in Charleville and Berlin, the Artillery Testing Commission decided upon a special gun of 9-cm to 10-cm calibre, high muzzle velocity and rapid rate of fire. The orders went to Krupp and Rheinmetall. In July that year both companies delivered designs for a trailer with motorized hauling vehicle. The basis for the 8.8-cm flak gun was the fast-loading L/45 on a pivoting chassis, work on which had been started in 1913 by the Imperial Navy. Rheinmetall delivered the first models for test purposes in 1917. The guns were satisfactory but the trailers had defects and were quickly discarded. Rheinmetall completed no further 8.8-cm flak of this type up to the war's end.

8.8-cm Flak 19 (Krupp)

(8.8-cm Flak L/45 on Motor-Towed Trailer (Krupp)

In April 1915 the Artillery Testing Commission set out the specifications for large calibre flak. The basis was Krupp's 8.8-cm fast-loading L/45 on central pivot mounting C/1913. Krupp delivered the first four prototypes in December 1916, by November 1918 a total of 169 had been delivered. Once the gun had been adopted it received the designation "8.8-cm K-Zugflak 19" (K-Zugflak= Krupp motor-drawn flak).

The gun was transported on a trailer drawn by Krupp-Daimler KD.1 heavy motor vehicle. A battery was composed of four such guns and four Flak-MG 08's.

Calibre:	88 mm
Barrel length:	3,960 mm= L/45
Weight for transport:	7,300 kg
Weight, firing position:	3,100 kg (without trailer)
Weight of shell:	9.6 kg
Traverse:	360°
Elevation:	0°/+70°
Muzzle velocity:	542–785 m/sec
Rate of fire:	10 rounds/min
Range, maximum:	9,800–10,800 m
Shell, maximum altitude:	5,200–6,850 m

10-cm Cannon 04

(Improvised Flak with rotatable upper base on fixed pedestal)

This was a gun of the Fussartillerie. Its ballistics made it suitable for engaging aerial targets, but its mounting was only built for ground targets. Shooting trials in 1909 and 1910 against aerial targets had proved the advantages of large-calibre flak on account of the superior explosive and splinter effect of the shells. During the First World War some of the few available guns of the type were put on pedestal mounts as flak.

Calibre:	105 mm
Barrle length:	3,130 mm = L/29.8
Weight, firing position:	2,755 kg
Weight of shell:	6 to 18.3 kg
Traverse:	360°
Elevation:	Up to +45°
Muzzle velocity:	424–580 m/sec
Rate of fire:	5 rounds/min
Range, maximum:	9,000–10,300 m
Shell, maximum altitude:	5,200–5,800 m

10 cm Gr. 15 mit Gr. Z. 04

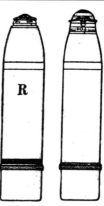

10 cm Gr. 15 mit Dopp. Z. 15 (umg.)

10.5-cm Flak L/35 (Rheinmetall)

Calibre:	105 mm
Barrel length:	3,699 mm = L/35
Weight, firing position:	2,710 kg
Weight of shell:	16 to 18.3 kg
Traverse:	360°
Elevation/depression:	–4°/+70°
Muzzle velocity:	461-580 m/sec
Rate of fire:	8 rounds/min
Range, maximum:	9,500–11,000 m
Shell, maximum altitude:	5,200–5,800 m

(On fixed pedestal)

The development of the 10.5-cm flak also stemmed from the request of the Artillery Testing Commission in April 1916. Krupp, Rheinmetall and Henschel were given the contracts. Subsequently heavy flak guns appeared on trailers hauled by motor vehicles, on railway waggons and on fixed pedestals. The latter were delivered as four models in June 1918. At the war's end the stock amounted to 36 which were used for air defence of the Reich.

3.7-cm Trench Gun (Krupp)

Calibre:	37 mm
Barrel length:	795.5 mm = L/21.5
Weight for transport:	(loaded on Army vehicle)
Weight, firing position:	203-223 kg
Weight of shell:	0.455 kg
Traverse:	45°
Elevation:	0°/+8°
Muzzle velocity:	400 m/sec
Rate of fire:	10 rounds/min
Range, maximum:	1,500 m

After the change to trench warfare the German infantry had at its disposal light mortars to provide plunging fire into enemy trenches and protected positions. The 3.7-cm gun was brought to the front for direct shooting from German trenches and to protect German assault troops. Krupp delivered this trench gun on an improvised mount of U-section iron and sheet metal plating with two 13-mm exchangeable splinter shields, a simple elevating mechanism and two spurs for anchoring the weapon. The barrel came from cannibalized 3.7-cm revolver guns.

The 3.7-cm trench gun (Krupp) came to the front in 1915 in the expectation that it would prove the right partner for the light mortar. However, the shell effect of this calibre fell short of requirements since now it was considered that an infantry shell should weigh 6 to 7 kg.

3.7-cm Pak (Anti-tank Gun) (Rheinmetall)

Calibre:	37 mm
Barrel length:	809 mm = L/21.8
Weight for transport:	465 kg
Weight, firing position:	175 kg
Weight of shell:	0.46 kg
Traverse:	21°
Elevation/depression:	–6°/+9°
Muzzle velocity:	506 m/sec
Range, maximum:	2,600 m

On the question of smaller guns to tackle tanks, not until the summer of 1918 did the Supreme Army Command decide on the calibre. Subsequently Krupp and Rheinmetall supplied four different 3.7-cm guns. Another design of the calibre came from an Oberstleutnant Fischer. Under the existing pressures of time, the Rheinmetall model with the rigid wheeled chassis made the best impression. Parts from the light mortar and its limber were incorporated into the design. For transport one horse or the four-man crew pulled the gun. The shell could penetrate 20 mm armour plate at 100 metres range. Of the 1,020 ordered, by November 1918 600 had arrived at the front. Every mortar batallion was to have 32 3.7-cm anti-tank guns of this type.

5.7-cm Rapid Fire Maxim-Nordenfeldt Chassis with Splinter Shield

Calibre:	57 mm
Barrel length:	1,504 mm = L/26.3
Weight for transport:	1,467 kg
Weight, firing position:	860 kg
Weight of shell:	2.73 kg
Elevation/depression:	−10°/+15°
Muzzle velocity:	401 m/sec
Rate of fire:	36 rounds/min
Range, maximum:	2,700–6,400 m

In 1887 the Belgian War Ministry decided to equip its army with a first series of 185 5.7-cm anti-assault guns on casemate chassis as offered by the British firm of Maxim-Norderfeldt. They used them to arm the fortresses at Namur and Liège amongst others. Later, the guns of this type became available on a wheeled chassis with splinter shield in the hope that they would prove effective between the forts. For the most part they were captured by the German Army in 1914. There were still 450 in German service at the front in 1916, used mainly as trench guns to support the infantry. In 1917 a number of the Belgian 5.7-cm guns were mounted on 4 tonne captured lorries and used as anti-tank guns. A small number were converted to guns for the assault panzer A.7.V.

6-cm Boat-Gun L/21 as Assault Support Gun

Calibre:	60 mm
Barrel length:	1,260 mm = L/21
Weight of shell:	.38-2.71 kg
Muzzle velocity:	437-448 m/sec
Rate of fire:	12-14 rounds/min
Range, maximum:	3,400 m

In 1900 the three naval batallions of the Imperial Navy received 6-cm boat guns L/21. They were mounted for naval infantry landing manouevres at the bow of a craft. In order that they should provide supporting fire after successful landings, each mother ship carrried wheeled chassis. In the autumn of 1914 the German Marinekorps in Flanders received four of these guns. They were held first by the Naval Artillery Regiment von Bernuth, and in mid-December 1914 went to II.Batallion/Naval Regiment as a boat-gun battery for direct support purposes. Here they served as a temporary replacement for infantry guns.

7.62-cm Infantry Gun L/16.5 (Krupp)

Calibre:	76.2 mm
Barrel length:	1,250 mm = L/16.5
Weight for transport:	1,146–1,156 kg
Weight, firing position:	608 kg
Weight of shell:	6 kg
Traverse:	9.4°
Elevation/depression:	-18.3°/+11.5°
Muzzle velocity:	295 m/sec
Rate of fire:	15 rounds/min
Range, maximum:	4,000 m

(Modified Russian Anti-assault Gun)

In 1914/1915 7.62-cm anti-assault guns were captured in Russian fortifications. When the frontline troops requested heavy support weapons in 1915, these captured weapons were remembered. Krupp received a contract to modify the guns and in the autumn of 1916 delivered them at first in small batches as infantry guns. The infantry-gun batteries of the appropriate Army assault sections were each given six 7.62-cm "Russian cannons" L/16.5 (Krupp). They could be moved about the battlefield either whole or in two parts by two or four horses respectively. They could also be pulled by the gun crew or towed by motor vehicle. Despite the great effect of a single round, guns rebuilt from captured material remained a makeshift solution. The short life of barrel and chassis was much criticized.

7.7-cm Infantry Gun L/19.5 (Rheinmetall)

Calibre:	77 mm
Barrel length:	1,510 mm = L/19.5
Weight for transport:	1,089 kg
Weight, firing position:	802–817 kg
Weight of shell:	6.55 kg
Traverse:	2.5°
Elevation/depression:	-10°/+15°
Muzzle velocity:	400–435 m/sec
Rate of fire:	15 rounds/min
Range, maximum:	4,500–5,100 m

In 1917 the War Ministry asked the Artillery Testing Commission to discuss with the major armament firms the manufacture of better infantry guns. At the beginning of 1918 Rheinmetall supplied the 7.7-cm infantry gun L/19.5. The 577 mm shorter barrel and breech were taken from the 7.7-cm field gun 96 n/A, the tail of the chassis was hinged. On the battlefield the gun could be transported in two sections or towed. It never arrived at the front, only infantry-gun battery No.5 at the Dourles military depot was equipped with it.

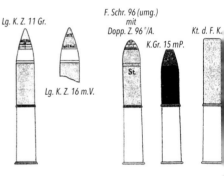

Lg. K. Z. 11 Gr.

Lg. K. Z. 16 m.V.

F. Schr. 96 (umg.) mit Dopp. Z. 96 °/A.

K.Gr. 15 mP.

Kt. d. F. K.

7.7-cm Infantry Gun L/27 (Krupp)

Calibre:	77 mm
Barrel length:	2,100 mm = L/27
Weight for transport:	1,045–1,150 kg
Weight, firing position:	845 kg
Weight of shell:	6.85 kg
Traverse:	6°
Elevation/depression:	–15°/+12°
Muzzle velocity:	465 m/sec
Rate of fire:	15 rounds/min
Range, maximum:	4,600–7,800 m

This Krupp gun was another makeshift. One-hundred-and-thirty-two of them were ordered at the beginning of 1917. In order to have heavy support weapons at the front as soon as possible for the assault batallions, barrel, chassis and barrel brake were taken unmodified from the 7.7-cm field gun 96 n/A: wheels and axle were new. The gun could be separated into two parts for transport, two each of four horses being teamed up for the purpose. It could also be pulled intact by the gun crew. Six guns were allotted to each infantry-gun battery. The original contract was reduced to forty-eight guns for eight batteries, these were formed in May and June 1918.

7.5-cm Mountain Gun L/14 (Krupp)

Calibre:	75 mm
Barrel length:	1,050 mm = L/14
Weight for transport:	1,010 kg
Weight, firing position:	550 kg
Weight of shell:	5.3 kg
Traverse:	5°
Elevation/depression:	-10°/+30°
Muzzle velocity:	300 m/sec
Rate of fire:	15 rounds/min
Range, maximum:	5,400 m

Before the outbreak of the First World War the German Army was not equipped with special guns for mountain warfare, although the colonial force in German South-West Africa had received some mountain guns from Rheinmetall in 1909. As both Rheinmetall and Krupp manufactured mountain guns for export they had some experience in this area. In August 1914 Krupp had four 7.5-cm mountain guns L/14 in store which were intended for the Chilean Army. These were purchased by the German Army administration and used in the winter of 1914/15 with other Krupp guns to set up nine mountain gun batteries (each of four guns). Transport was horse-drawn with a limber or the gun was broken down into seven sections for carriage by pack animals.

7.5-cm Mountain Gun M.15 (Skoda)

Calibre:	75 mm
Barrel length:	1,155 mm = L/15
Weight for transport:	766.1 kg
Weight, firing position:	620 kg
Weight of shell:	6.5 kg
Traverse:	7°
Elevation/depression:	-9°/+50°
Muzzle velocity:	224-350 m/sec
Rate of fire:	8 rounds/min
Range, maximum:	6,650-7,000 m

At the end of 1916 the War Ministry decided to re-arm the mountain artillery of the German Army with the 7.5-cm mountain gun M.15 (Skoda) which had been adopted in Austria. The idea of a uniform armament for the mountain artillery of both countries was decisive although the Skoda model was also very accurate and had longer range than comparable German guns. The first batteries with this type of gun arrived at the front in the spring of 1917. One year later the mountain artillery had eight batteries of four such guns each, while the German Army also used this type as an infantry gun. In May and June 1918 the War Ministry prepared to form fourteen infantry gun batteries, but in this arrangement their use proved unsatisfactory: they were built to be carried by pack animals, not for towing over long distances, the wheels being too small and narrow for the purpose.

10.5-cm Mountain Howitzer M.1912 L/12 (Rheinmetall)

Calibre:	105 mm
Barrel length:	1276 mm = L/12
Weight for transport:	1,095 kg, divisible into eight sections for carriage by pack animals, 1,215 kg with limber
Weight, firing position:	815 kg
Weight of shell:	15.8 kg
Traverse:	6°
Elevation/depression:	-7°/+40°
Muzzle velocity:	256 m/sec
Rate of fire:	2 rounds/min
Range, maximum:	5,200 m

Trench warfare in mountainous terrain gave rise to demands for greater effect from single rounds of fire, and for engaging targets behind steep cover. A large-calibre mountain howitzer with various propellants was what was looked for. Krupp and Rheinmetall took up the development of these guns. While Krupp delivered 36 of the 10.5-cm mountain gun L/11 in 1916, Rheinmetall supplied only one model. The guns could be dismantled into eight parts for transport by pack animal but also be drawn by mules with the limber, using bar or dual shafts. The German designs for mountain howitzers proved inferior in performance particularly to Austrian guns.

10-cm Mountain Howitzer M.16 L/19 (Skoda)

Calibre:	105 mm
Barrel length:	1,930 mm = L/19
Weight for transport:	2,090 kg
Weight, firing position:	1,235 kg
Weight of shell:	13.5–16 kg
Traverse:	5.5°
Elevation/depression:	–8°/+70°
Muzzle velocity:	180–325 m/sec
Rate of fire:	4 rounds/min
Range, maximum:	7,750 m

Because German mountain howitzers transported by pack animals were inferior in ballistic performance to their Austrian mobile counterparts, the War Ministry negotiated with the Austrian Army administration to take four 10-cm mountain howitzers M.16 (Skoda). Skoda delivered them with the German 10.5-cm calibre so as to be able to fire German howitzer ammunition later. The battery appeared at the front in May 1917 where despite the unaccustomed

positional changes in the mountains (for which they had to be broken down into three sections) they proved their worth. These also met all requirements for range and shell effect.

3.7-cm Revolver Gun (Hotchkiss System)

Calibre:	37 mm
Barrel length:	1,192 mm = L/32
Weight, firing position:	571 kg (on casemate chassis)
Weight of shell:	0.46–0.51 kg
Traverse:	20°
Elevation/dpression:	–10°/+15°
Muzzle velocity:	400 m/sec
Rate of fire:	40–50 rounds/min
Range, maximum:	2,000–3,000 m

The Artillery Testing Commission had been conducting tests on the French revolver-gun since 1880. It was accepted in 1885 after years of testing and manufactured in series at the Gruson Works, Magdeburg. In the Army it was a weapon of the Fussartillerie which used it as a flanking gun in fortifications when the bundle of five barrels would be arranged by bolting them to a firing stand through a retaining plate so that fire could be *kaponniert* (see Glossary) along the fortification trenches. A gunner turned the barrels with a crank while two others loaded shells into the filler shafts. A few 3.7-cm revolver guns were mounted on adapted field-gun chassis C/73 while the Navy used them aboard unprotected cruisers and gunboats, also on the 8-cm boat chassis.

In the First World War the 500 weapons stored in depots with their ammunition were suddenly remembered, and a number arrived at the front and were used in the trenches to ward off infantry assaults, others were converted into flak. Finally, the barrels were used for infantry and anti-tank guns.

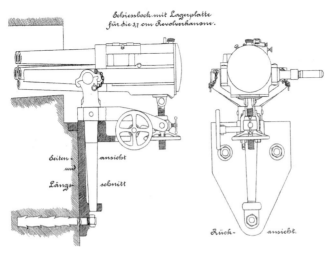

Schiessbock mit Lagerplatte
für die 3,7 cm Revolverkanone.

Seiten- und Längs- ansicht schnitt

Rück- ansicht.

5-cm Cannon on Casemate Chassis

Calibre:	53 mm
Barrel length:	1,300 mm = L/24.5
Weight, firing position:	566 kg
Weight of shell:	1.66–2.38 kg
Elevation/depression:	–10°/+15°
Muzzle velocity:	457 m/sec
Rate of fire:	20–25 rounds/min
Range, maximum:	3,000 m

The Magdeburg firm Gruson had achieved a significant improvement in quick-firing with their single barrel gun design. In 1888 the Artillery Testing Commission began trials with the Gruson 5.3-cm gun which proved itself to be so superior to the 3.7-cm revolver cannon that in 1890 orders went out to adopt the former. One-hundred-and-thirty-two of these were delivered on the casemate chassis. In this format the guns were used only in flanking installations of the fortresses at Metz and Istein. Their purpose was defence against close range attack on fortified positions using shrapnel (Kartätschfeuer – see Glossary). In the First World War they were retained by the Fussartillerie as antiquated fortress-guns.

5-cm Cannon on Chassis within Armoured Cupola

(Gruson Mobile Panzer)

Calibre:	53 mm
Barrel length:	1,300 mm = L/24.5
Weight for transport:	2,940 kg
Weight, firing position:	2,240 kg
Weight of shell:	1.67 kg
Traverse:	360°
Elevation/depression:	-3°/+10°
Muzzle velocity:	457 m/sec
Rate of fire:	30-35 rounds/min
Range, maximum:	3,000 m

This mobile gun within armour arose from collaboration between Schumann and Gruson. It was tried out on the artillery testing range at Kummersdorf in 1888 and over the next two years used to equip the fortresses. Army administration approved 190 of this type. Their purpose was to reinforce and complement infantry fire in fortifications and fortified areas. The armoured cupola was of 40 mm ingot steel, the lower jacket 4 mm thick, 15 mm to the rear. The chassis had space for 120 shells. The whole unit was mobile with a single-axle limber pulled by three horses. In order to be located for firing, a segment of twin track of 575 mm gauge was carried. The gun was ready to fire in two hours. In the First World War 5-cm cannons on chassis within armoured cupola served at the front as anti-assault guns.

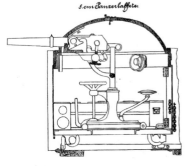

5.cm Panzerlaffete.

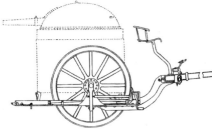

Aufgeprotzte 5cm Kanone in Panzerlaffete.

10-cm Armoured Turret, Reinforced

At the beginning of 1893 the Artillery Testing Commission received a contract from the War Ministry to examine the fitness for purpose of a 12-cm cannon. As the new barrel length for a fortress gun, the 10.5-cm calibre already in use by the Navy was suggested. Shielded pedestals and armoured turrets were to be included in the designs. The 10-cm armoured turret with a 150-mm armoured ceiling was introduced in 1900. The procurement cost for a 10-cm armoured turret battery of 4 guns, 4 reserve barrels and 3000 shells ran to 2,364 million marks. Because of the expense, the quantity of 10-cm armoured turrets in the first-line artillery array in the important fortresses at Metz and Diedenhofen was small.

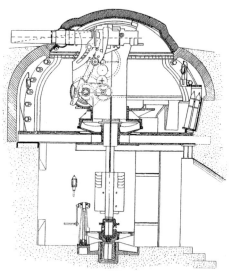

Calibre:	105 mm
Barrel length:	3,680 mm = L/35
Weight, firing position:	94,250 kg (armoured turret with gun barrel and above-ground armour)
Weight of shell:	16.06–18.75 kg
Traverse:	360°
Elevation/depression:	–5°/+35°
Muzzle velocity:	585 m/sec
Range, maximum:	8,500–10,800 m

10-cm Cannon with Shielded Mount

Calibre:	105 mm
Barrel length:	3,680 mm = L/35
Weight, firing position:	17,550 kg
Weight of shell:	16.06–18.75 kg
Traverse:	150–160° (360°)
Elevation/depression:	–5°/+35°
Muzzle velocity:	585 m/sec
Rate of fire:	9 rounds/min
Range, maximum:	10,800–13,100 m

As an alternative to the large scale and costly armoured batteries, in 1897 10-cm fixed assault batteries were installed in German fortresses. The guns were embedded on a concrete base rotatable by means of a pivoting pedestal. An 80 mm thick armour cupola protected the gun against enemy fire. There were batteries with 3, 4 or 6 10-cm cannons on a shielded base. Fifty-eight units were manufactured and located in the fortresses at Graudenz, Metz, Molsheim, Neubreisach, Strasbourg and Thorn.

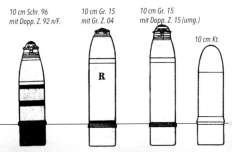

10 cm Schr. 96 mit Dopp. Z. 92 n/F.   10 cm Gr. 15 mit Gr. Z. 04   10 cm Gr. 15 mit Dopp. Z. 15 (umg.)   10 cm Kt.

15-cm Turret Howitzer

These guns had already been introduced into service as early as 1893 and in the fortresses were first-line guns. They were built in batteries of four in the fortifications at Strasbourg and Metz (here amongst others two at Fort Lothringen). Their ballistic performance excelled that of the 15-cm heavy field howitzer.

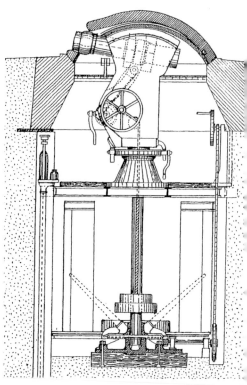

Calibre:	149.7 mm
Barrel length:	1,720 mm = L/11.5
Weight, firing position:	83 tonnes
Weight of shell:	42.3 kg
Traverse:	360°
Elevation:	0°/+40°
Muzzle velocity:	276–329 m/sec
Rate of fire:	2 rounds/min
Range, maximal:	6,050–7,200 m

8.8-cm Fast-Load Cannon L/45

On centre-pivoting mount C/1913 (Naval Flak Krupp)

Krupp had completed the design work on this gun before the First World War. It had been well tested and had been accepted by the Imperial Navy. In 1915 the first guns were delivered to Flanders where in fixed emplacements they achieved very good results in coastal defence and anti-aircraft work. The Navy made one 8.8-cm fast-load cannon L/45 available to the Army administration as a model for the development of its own flak.

Calibre:	88 mm
Barrel length:	3,960 mm = L/45
Weight, firing position:	6,060 kg (with bedplate and protective armour)
Weight of shell:	9.6 kg
Traverse:	360°
Elevation/depression:	–10°/+70°
Muzzle velocity:	750 m/sec
Rate of fire:	15–20 rounds/min
Range, maximum:	13,300 m
Shell, maximum altitude:	c. 7,500 m

10.5-cm Fast-Load Cannon L/45

On centre-pivoting mount C/1913

These guns came from the small protected cruisers. Set on a bed of concrete or iron in fixed emplacements they were used for coastal defence and flak. During the First World War four-gun batteries were operational in Flanders, amongst other locations at Knocke on the Dutch border, at Zeebrugge and Raversijde.

Calibre:	105 mm
Barrel length:	4,725 mm = L/45
Weight, firing position:	5,520 kg (without bedplate)
Weight of shell:	17.4 kg
Traverse:	360°
Elevation/depression:	0°/+70°
Muzzle velocity:	720 m/sec
Rate of fire:	5 rounds/min
Range, maximum:	13,900 m
Shell, maximum altitude:	7350 m

15-cm Fast-Load Cannon L/40

On revolving pedestal

Originally these 15-cm fast-load cannons L/40 were fitted aboard the Wittelsbach-Class pre-Dreadnought battleships laid down in 1901-1902. A number of these obsolete guns were mounted on a revolving pedestal and given a new role during the First World War with the Army Fussartillerie. In 1915 some were operational for the coastal defence of Flanders, amongst others, four guns protected the harbour mole at Zeebrugge and another four were in the dunes at Blankenberg, Bredene and Raversijde.

Calibre:	149.3 mm
Barrel length:	5,960 mm = L/40
Weight, firing position:	11,500 kg
Weight of shell:	44 kg
Traverse:	120°
Elevation/depression:	–8°/+32°
Muzzle velocity:	750 m/sec
Rate of fire:	4 rounds/min
Range, maximum:	18,700–19,00 m

17-cm Fast-Load Cannon L/40

On central pivoting platform with rotatable hood

Guns of this configuration were brought out from naval stocks from the spring of 1917 and delivered to the Flanders coast. They originated from the pre-Dreadnought battleships of the "Deutschland" and "Braunschweig" classes built between 1902 and 1906. The guns, some given a protective hood, were present in batteries of two or four guns at Knocke, Breedene and Ostende on the Belgian coast.

Calibre:	172.65 mm
Barrel length:	6,900 mm = L/40
Weight of shell:	63 kg
Traverse:	360°
Elevation:	0°/+45°
Muzzle velocity:	815 m/sec
Rate of fire:	1 round/min
Range, maximum:	24 kms.

17 cm Spgr. L/4,7 (Kz.)
(Haube)
mit Kz. f. Spgr. m. K. u. St.

Schussfertig!

28-cm Coastal Howitzer L/12

In coastal-howitzer (central pivot) emplacement

In 1908 the 28-cm coastal howitzer was delivered to the coastal artillery. The 345 kg AP shell could pass through 171 mm thick armour plate in 63° plunging fire. Armour decks 100 mm thick offered no protection at a range of 9,900 metres. The howitzer and its crew had a 60 mm armoured shield for protection against flying fragments and fire from light weapons.

The gun was operational in coastal-mortar batteries of from four to eight guns. In 1914 on the island of Borkum there were two four-gun batteries, at Pillau and Cuxhaven one four-gun battery each and on Heligoland one eight-gun battery. The four-gun battery at Wilhelmshaven was transferred to Blankenberge in Flanders. After 1907 Danzig also had a battery with four 28-cm howitzers. On the Chinese mainland, the German colony at Tsingtau had four guns of this type facing out to sea.

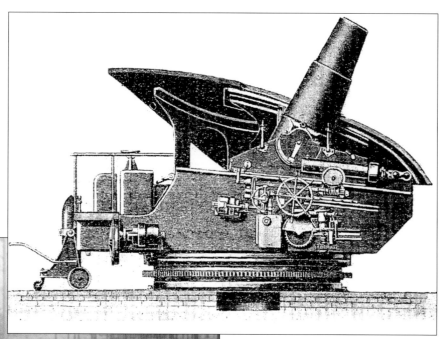

Calibre:	283 mm
Barrel length:	3,400 mm = L/12
Weight, firing position:	41,100–63,300 kg
Weight of shell:	215–345 kg
Traverse:	360°
Elevation:	0°/+65°
Muzzle velocity:	379–425 m/sec
Rate of fire:	1 round/min
Range, maximum:	9,900–11,400 m

28-cm Coastal Cannon L/21

On central pivot emplacement (Russian 11-inch coastal cannon M.1877)

Until about 1880 the Russians purchased their coastal artillery and ship's guns almost exclusively from Krupp. The 11-inch coastal cannon L/21 was amongst them. During the First World War the German Army captured a number of guns of this type and put them to use, some with ammunition manufactured in Germany. In 1916 four of the guns were operational: in 1918 all were retired from service.

Calibre:	280 mm
Barrel length:	6,104 mm = L/21.8
Weight, firing position:	57,400 kg
Weight of shell:	213.1 kg, 182.2–250 kg (Russian original ammunition)
Traverse:	120°
Elevation/depression:	–6°/+20°
Muzzle velocity:	518.5 m/sec
Rate of fire:	1 round every three minutes
Range:	12,500–14,400 m (with German ammunition)

28-cm Ring Cannon L/22

(On central pivot framework)

Calibre:	283 mm
Barrel length:	6,226 mm = L/22
Weight, firing position:	50,778 kg
Weight of shell:	250 kg
Traverse:	120°
Elevation/depression:	–6°/+20.5°
Muzzle velocity:	457 m/sec
Rate of fire:	1 round every three minutes
Range, maximum:	6,400–10,600 m

In 1878 Krupp delivered the first coastal guns of this type to Fort Kugelbake near Cuxhaven in the Elbe estuary. A battery with a total of ten 28-cm ring cannons L/22 was installed. By 1906 the gun was numbered amongst the older coastal artillery weapons possessed by the Imperial Navy since 1888. In the spring of 1915 some of them were shipped to Flanders in the vicinity of Zeebrugge and they were not replaced until 1917 by 17-cm fast-load cannons with a range of 21,300 metres.

38-cm Fast-Load Cannon L/45

(Embedded in firing emplacement with or without protective hood)

Calibre:	380 mm
Barrel length:	17,100 mm = L/45
Weight, firing position:	270 tonnes
Weight of shell:	400–800 kg
Traverse:	144°
Elevation:	Up to +45°, up to +55°
Muzzle velocity:	750 m/sec
Rate of fire:	1 round every 4 to 10 minutes (in the traverse field from 0° to 10°
Range, maximum:	24,000–47,500 m

In the autumn of 1914, Supreme Army Command ordered six firing emplacements for 38-cm fast-loading cannons. When work began on the Western Front it was seen that the time required for installation would be too great. Therefore at the end of 1916 three guns were taken to Flanders to strengthen the coastal defences. In 1916 Krupp received another order for four 38-cm emplacements with 50 mm armour plating for protection. Towards the end of the war the "Pommern" battery in the park at Moere had one such gun and the "Deutschland" battery between Breedene and Klemskerke (1. Long Range Group Ostend) had four of this type.

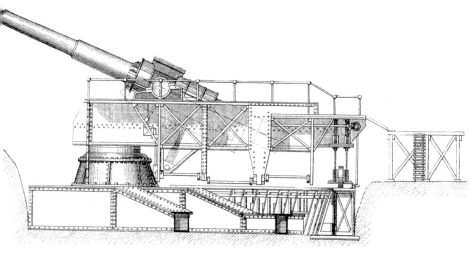

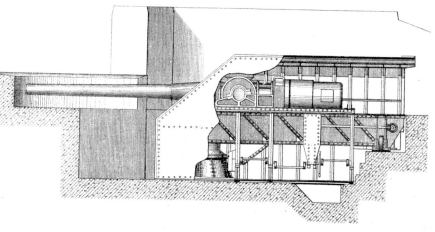

Anti-balloon Guns (Ballonabwehrkanone B.A.K)
A gun designed for use against observation balloons, airships and aircraft. Even before the First World War a centre-pivoting gun installed on the rear of truck, trailer or other stable-chassis had found favour. The important features were all-round traverse and large degree of elevation, special ranging devices and a fast rate of fire. The term was later replaced in German by "Flak" (Flugzeugabwehrkanone).

Anti-tank Gun (Tankabwehrkanone TAK)
First World War term. The TAK was a small-calibre gun built specially to engage tanks. The term was later replaced in German by PAK (Panzerabwehrkanone).

Artillery Testing Commission (Artillerie-Prüfungs-Kommission A.P.K.)
From 1890 this independent authority was answerable directly to the War Ministry. It had three departments, (1) Field-artillery, (2) Fussartillerie and Testing (at the Kummersdorf ranges), (3) Depot administration. Amongst its aims were the development of artillery and materials, the testing and evaluation of the artillery of foreign armies, to remain informed as to technological advances in artillery and to use the results for the further development of the weapons branch. Inventions were assessed, reports drafted, guns and other artillery equipment tried and investigated.

Barrel brake, barrel recoil
After a shell is fired, the gun barrel recoils along the barrel cradle until intercepted by the barrel brake. A powerful spring then returns the barrel to the firing position.

Calibre
Expression of the barrel length as related to the calibre. For example: 7.7-cm experimental field gun L/35 (Krupp). To calculate length of barrel: L=35 x 77 mm =2,695 mm

C-Shell (C-Geschoss)
A long-range shell for the field artillery introduced in 1917. It had a slim head and tapered towards its base. The C-shell for the 7.7-cm field cannon 16 weighed 5.89 kg of which 0.55 kg was HE. With a muzzle velocity of 602 m/sec such a shell would impact at 10,700 metres, a gain in distance of 1,600 metres over the standard shell.

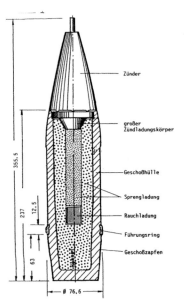

Zünder

großer
Zündladungskörper

Geschoßhülle

Sprengladung

Rauchladung

Führungsring

Geschoßzapfen

355,5

237

12,5

63

Ø 76,6

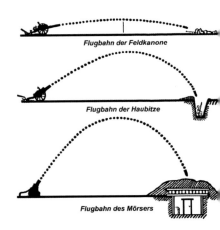

Flugbahn der Feldkanone

Flugbahn der Haubitze

Flugbahn des Mörsers

Einheitsmunition

In 1914 the German field artillery introduced the 7.7-cm field gun shell 11 and the 10.5-cm field howitzer shell 05. Both could be used with explosives or shrapnel. This corresponded to the prevailing opinion before the war regarding the types of ammunition which the field artillery should have, simplified management and matters of ammunition supply. Shell 11 contained 294 hard lead balls each of 10 grams, 0.25 kg explosive and an ignition charge of 0.08kg at the base of the shell. The disadvantages were the limited proportion of explosive and the complicated structure of the shell. These shells could not take Granatfüllung 88 (q.v.)

Flight trajectory of shells

In the First World War modern guns were classified by their location of purpose (e.g. mountain guns) or their calibre (light, medium and heavy). They were also identified by their shell trajectory. Field guns were flat trajectory and mortars steep trajectory weapons. The howitzer fitted between these two. In the German Army the use of the terms flat trajectory-and steep trajectory-guns was common.

Fussartillerie

In earlier times the mobile batteries of the field artillery, as opposed to the mounted artillery, had been designated "foot artillery". In the German Army, in 1872 the fortress artillery, fully separated from the field artillery, was now designated "foot artillery". Its weaponry, equipment and organisation turned it into the heavy artillery of the field army. There were regiments and reserve regiments of foot artillery as well as independent foot artillery batteries.

Haubengranate

Shells with a hollow cap above the fuze in order to achieve especially long distance.

Kanonengranate 15 (mit Panzerkopf) (Cannon-shell 15 with AP head)

Armour-piercing shell for 7.7-cm field guns. It was fabricated from slugs for cannon shell 15 and given an armour head. Below this head was an impact fuze

with delay. The shell would penetrate 40 mm armour at 90° at 2,000 metres.

Kaponniere

Kaponniere were structures built inwards from a fortress wall to house Kartätsch guns (q.v.) including revolver cannons sited at right angles to the front for the purpose of offering flanking fire to attackers penetrating the fortification trenches.

Kartätschen

These were zinc or tin boxes filled with zinc or lead-antimony balls and a propellant. A Kartätsch for the heavy field cannon C/73/91 for example contained 76 balls. By the First World War the weapon had diminished in importance but was then reintroduced for various gun types.

Kartaune

In earlier times, the common designation for a gun which fired a cannon ball only a quarter the usual weight. The term fell into disuse at the beginning of the 19th century but reappeared during the First World War. Then it came to mean, for example, a shell of 333 kg fired from a 30.5-cm barrel on the 42-cm mortar chassis, while the 42-cm mortar fired a shell of 810 kg.

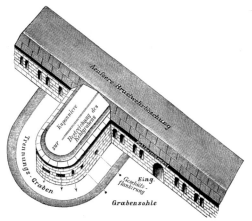

A	Kartätschbüchse mit Zinkeinlage
B	Kartätschbüchse mit Stahlblecheinlage
D	Schlußscheibe
H	Holzeinsatz
S	Sicke zum Zerreißen der Büchse

b	Umfederung der Büchse
c	Kupferband zum Zerreißen der Büchse
d	Öse
e	Strickhandgriff

Shrapnel

Named after its inventor, Englishman Henry Shrapnel (1761-1842), this was a thin-walled artillery shell filled with an explosive charge and hard lead or steel balls. A time fuze would expel the balls in the desired direction for effect against ground troops. As an example, the shrapnel shell from a 9-cm cannon had 262 balls each of 13 grams of weight, the 15-cm cannon had 633 balls each of 28 grams of weight.

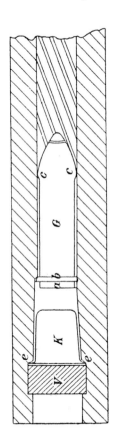

Load chamber

The part of the gun barrel which houses the shell and cartridge. After detonation and ignition of the propellant charge, the gas pressure necessary to expel the round develops: the length of the loading chamber influences the muzzle velocity of the shell.

Shell-filling 88 (Granatfüllung (88)

A yellow explosive composed of carbolic acid nitrate (picric acid, trinitrophenol). Very high velocity at detonation (7,250 m/sec) and temperature (3,230°C) in combination with metals creates shock-sensitive salts (picrates). Granatfüllung 88 was used as an explosive in field-artillery and Fussartillerie shells (except for Einheitsmunition (q.v.) and shrapnel shells.)

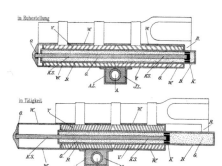

in Ruhestellung

in Tätigkeit

A Geschosskern
a walzenförmiger Teil
B Stoßboden
C Füllkugeln
c Bogenspitze
F Führungsring
K Kammerhülse
L Bodenkammer
M Mundloch
W Zentrierwulst
Z Zünder

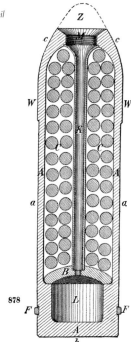

878

A. Achse
A.L. Äußeres Lager
B. Bremszylinder
G. Mischung aus
    Glyzerin und Wasser
J.L. Inneres Lager
K. Kolben

K.S. Kolbenstange
R. Rohrhalter
O. Befestigung
    der Kolbenstange
    an der Oberlafette
V. Vorhaltfeder
W. Wiege

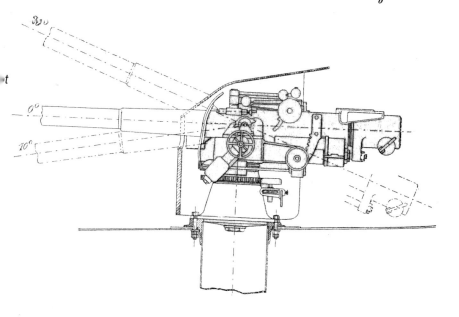